# 名宅生活美学
## 时尚与舒适的空间对话

Well-known House Life Aesthetics:
Space Dialogue Between Fashion and Comfort

精品文化 编

华中科技大学出版社
http://www.hustp.com

中国·武汉

# Preface

Residence design is a very difficult task. Designer needs to not only satisfy his own desire on design, but also satisfy the clients—family member's design. The clients have different ages, personalities and individual preferences, etc.. Unfortunately, a good design usually needs a consistent and clear theme, so that most of the award-winning residential projects are suitable for young couples and singles for those works have more stories and coherence.

The book is named Well-known House Life Aesthetics. In fact, the definition of the well-known house is very vague. Does it a celebrities' house or a famous architect's house? Does representative celebrity know what is life aesthetics? Does the one able to live in the well-known house know live aesthetics? I think life itself is an art, one should know how to appreciate beauty, enjoy beauty, and praise beauty, then the gain of personal virtues and qualities means understand life aesthetics. Among my works of residential projects, the majority is designed for well-tasted clients. Their common points are that they have been under the study of western education, and they traveled thousands of miles from an early age, and have lots of knowledge accumulation. And they know how to find their own proper designer, and they show respect to the design work. As a good designer, things that more important than the design experience is the experience of life. We shall pay more requirements on our own life rather than the clients, and try to live better than they live. Those experiences are not necessarily realized through our own money, for many times we learn the experience from different backgrounds' clients. I have a pet phrase: "If one has not once lived in the five-star hotel, one can not design a five-star hotel.".

Design is actually a psychology, but it does not require to be theoretical. We should understand the client's psychology through constant listen and analysis; we should understand the problem and use talent and exercised aesthetic to meet client's needs, which is a win-win good design.

I am recently on behalf of Hong Kong Interior Design Association to attend "East Gather" conversation hold by Hong Kong, Japan, South Korea's Joint Design Association. During the activity, I saw the difference on different cultures in the design and the fade and changes under the change of the economic system. Basically, different country has different definition and cognition on "luxury" and "simple". Japanese simplicity maybe seems lack of vitality to the Koreans and Chinese people. However, the Chinese luxury maybe too cluttered and lack of humane to the Japanese. I remember once in the exchange activity, after the speech of a Hong Kong designer, a Japanese teacher said: "Your design is really good, but I have a little lost, because from the work I saw China's rise and Japan's retrogression in the decade. In the past, we were not need to consider the costs problem." I find the same phenomenon in many clients. In many times, clients use the high price projects, but not base on their own design requirements.

As a Hong Kong designer, I think Hong Kong is located at a mixed cultural center point. We have a rich worldview which is our advantages. I personally do not like to discuss design trends, especially in residential design. Because these are very personal things, these reflects a unique taste and need, rather than pursuing the trend. I can only say that my favorite traditional European design may slowly be eliminated through change of time and world view. In the past, we pursue the European-style residence as a luxury status symbol, but not a true aesthetic taste. With China's booming, I believe people will soon know how to pursue their own taste, rather than blindly chasing the trend.

All texts are directly translated from the Chinese version. Only the chinese text is originally written by Wesley Liu.

Wesley Liu
Design Director
HKIDA (Hong Kong Interior Design Association) Executive Committee Member
Chairman of APIDA (Asia Pacific Interior Design Awards) 2013-2015
Chairman of East Gathering Committee

住宅设计是一门非常高难度的工作。设计师不但要满足于自己对设计的欲望，还要满足客户家庭成员的需求。他们有着不同的年龄、性格和个别喜好等等。遗憾的是，一个好的设计通常都需要有连贯性和明确的主题，因而大部分获奖的住宅项目一般适用于年轻夫妇和独身的客户。因为这样出来的作品会比较有故事性和有连贯性。

这本书叫《名宅生活美学》。其实名宅的定义是非常模糊的，到底是名人住的住宅还是著名建筑师盖的住宅呢？有代表性的名人就懂得什么是生活美学吗？或是说有能力住进名宅就会懂得生活美学吗？我觉得，生活本身就是一门艺术，懂得去欣赏美、享受美、称赞美，从而得到个人品德和素质方面的培养才是叫作懂得生活美学。在我做过的住宅项目当中，大部分都是生活有品位的客户。他们的共通点都是受西方的教育，从小就走过万里路，见识广博，而且他们懂得去找适合自己的设计师，并尊重设计作品。作为一个好的设计师，设计经验固然重要，但我觉得最重要的是生活的经验，我们要对自己的生活比客户更有要求，活得要比他们更精彩。这些经验不一定全部都要自己用钱去体验，很多时候我们是从不同背景的客户身上学到了他们的经验。我还有一句嘴边话："没住过五星级酒店，又怎能去设计五星级酒店呢？"。

设计其实也是一门心理学，但不需要理论化。我们要通过不断地聆听和分析去理解客户的心理，懂得解决问题并用天赋和锻炼出来的美感来满足客户的需求，这才是一个双赢的好设计。

我近年来代表香港室内设计协会，在香港、日本、韩国三地设计协会合办的"东方聚合"交流活动之中也看到了不同文化在设计上的差异，以及随着经济体系的变化在设计上的蜕变。基本上各地对"豪华"和"简约"的定义和认知都有很大的分歧。日本人崇尚的简约，对于韩国人和中国人而言可能会显得缺乏生息。然而中国人的豪华可能对日本人来说会显得太杂乱和不够人性化。我记得有一次在三地交流活动中，有一位日本老师看完一位香港设计师的演讲后说了这样一句话，"你的设计真的很棒，但从中我也看到有一点失落。因为从作品中我看到了这十年间中国的崛起和日本的倒退。以前我们的设计都不太需要考虑到成本问题的。"我在众多客户中也发现了这一点，很多时候客户会因为价格高而采用，而不是因为设计需要的本身。

作为一个香港的设计师，我觉得我们是一个文化交杂的中心点，有着丰富世界观的优势。设计趋势方面我个人不喜欢去讨论，尤其在住宅设计方面。因为这些都是很个人化的事，反映了一种独特的品位和需求，而不应该是追求趋势的。我只能说现在喜欢的传统欧式设计慢慢应该会随着年代和世界观的变化而逐渐被淘汰。以前我们住宅追求的欧式风格是一种豪华身份的象征，但不是一种真心的美学品位。随着中国的蓬勃发展，我相信大家很快会懂得如何追求自己的品味，而不是盲目追赶潮流。

廖奕权
设计总监
香港室内设计协会委员会理事
2015 APIDA 亚太区室内设计大奖筹委会主席
东方聚合筹委会主席

# Contents

| | | |
|---|---|---|
| 006 | 金地澜菲溪岸E1户型 | Jindi Lanfei River Bank EI Unit |
| 016 | 观湖壹号（一） | Lake Enjoyment No.I(I) |
| 022 | 树影下之悠然 | Mandarin Oriental Apartments |
| 032 | 泷景复式户型 | Oriental Bund Duplex Unit |
| 038 | 清雅灵韵　内敛悠长 | Quite and Elegant Charm with Distant Connotation |
| 046 | 沙田独立屋 | House in Shatin |
| 058 | 红树西岸花园 | Mangrove West Coast |
| 064 | 北角赛西湖 | North Point Breamar Hill Mansion |
| 070 | 薄扶林贝沙湾 | Pok Fulin Residence Bel-air |
| 078 | 自然意象 | Deep in Nature |
| 090 | 世界花园 | Worldwide Garden |
| 098 | 帝景台（二） | Dynasty Heights II |
| 106 | 观湖壹号（二） | Lake Enjoymen No.I(II) |
| 112 | 恒懋大楼 | Hannover Court |
| 124 | 皇璧 | The Westminster Terrace |

目录

| | | | |
|---|---|---|---|
| 128 | 康乐园 Hong Lok Yuen | 230 | 瑧环A户型 Gramercy A Unit |
| 140 | 匡湖居 Marina Cove | 234 | 帝峰皇殿 Hermitage |
| 148 | 岚岸 Sausalito | 240 | 毕架山花园 Beacon Heights |
| 156 | 泷景B户型 Oriental Bund B Unit | 246 | 比华利山别墅 The Beverly Hills |
| 162 | 名家汇 Hill Paramount | 256 | 半山壹号 The Joyful Tree House |
| 168 | 葡萄园 The Vineyard | 262 | 嘉富丽苑 Clovelly Court |
| 176 | 气转乾坤 CHI on the Jardine's Lookout | 270 | 海逸豪园 Laguna Verde Tower |
| 184 | 溱岸8号 The Rivapark | 278 | 鲗鱼涌示范单位 Quarry Bay Showflat |
| 188 | 青榕台 Tsing Yung Terrace | 286 | 御凯 The Dynasty |
| 196 | 擎天半岛 Sorrento | 292 | 雅士花园 The Astoria Tower |
| 202 | 颐和湖光山舍 Yihe Mountain & Lake House | 302 | 西贡匡湖居 Marina Cove |
| 206 | 山顶大宅 House on the Peak | 310 | 涤涛山 Constellation Cove |
| 210 | 深井碧堤半岛 Bellagio | 320 | 大埔鹿茵山庄 Tai Po Deerhill Bay |
| 218 | 天御豪庭B户型 Type B Apatment La Couronne | 326 | 帝景台（一） Dynasty Heights I |
| 224 | 升御门 Chatham Gate | | |

# Jindi Lanfei River Bank E1 Unit

金地澜菲溪岸 E1 户型

设计单位：北京风合睦晨空间设计
设 计 师：陈贻、张睦晨
项目面积：224 ㎡
项目地点：湖北武汉
主要材料：月桂银灰石材、木地板、壁纸、白色乳胶漆

Design Company: Beijing Fenghe Muchen Space Design
Designers: Chen Yi, Zhang Muchen
Project Area: 224 m²
Project Location: Wuhan, Hubei
Major Materials: Laurel silver gray stone, Wood floors, Wallpaper, White paint

随着现代社会中人们的生活品质不断提高，人们对于找寻内心精神体验及自我定位的愿望变得越来越强烈，空间的意义已经完全超越了现实生活实际物质需求的层面，而被要求达到更高的精神诉求层面。更多的人及设计师关注和考虑的是，如何让空间跳出具象的物质属性层面而达到空间抽象精神层面，从而使空间突显出独特的设计气质与艺术品位。

这一次，设计师呈现给我们的是一处优雅、低调、时尚、新颖的居室空间。项目虽然以欧式风格为主，然而整体的空间却散发出一种独特的现代艺术魅力。现代平面构成的视觉形式语言被巧妙地融入到空间中，让使用者可以在优雅、美好的艺术气息中享受温暖阳光下的蓝调生活，体会属于自己的静谧时光。

经过设计师的反复推敲，对空间布局的精心规划，原有户型的功能劣势得以全然消除。简化后的现代形式语言和传统、丰富的线性形式语言形成了强烈的视觉反差及对比。设计师在挑空空间中巧妙地增设了

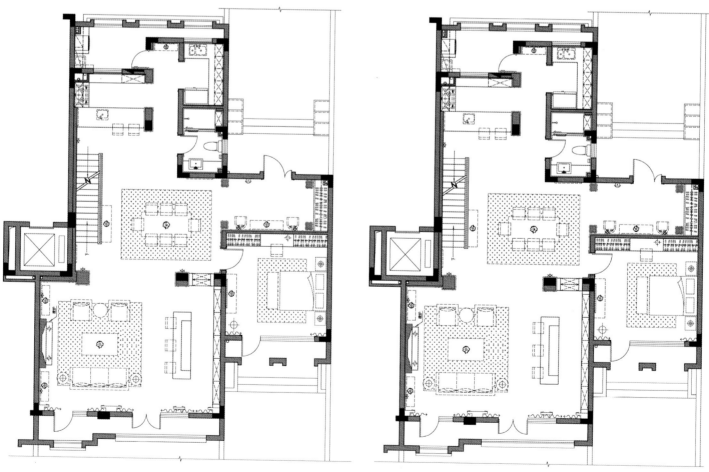

二层的连廊部分,将纵向空间进行更为合理地划分,从而形成了更为多变的空间感受和功能区域划分。

内敛、低调的浅灰色被大胆运用,作为大面积的背景底色,并辅以净透、跳跃的白色为点缀,营造出一种明快、简约的空间感受。总体氛围融入了时尚、舒适内敛的设计元素,突出居住者的沉着硬朗、睿智深刻的生活阅历及艺术品位,展现出空间的立体感受和特有的文化品位,同时也满足了居住者对生活品质及舒适居所的完美追求。

With the development of modern society life quality, people have strong willing to seek for inner spiritual experience and self position. Space meaning is completely over the actual material needs of real-life level. Life is required to achieve higher spiritual level. People and designers pay more attention on how to make material space jump to the abstract spiritual space, making the space has unique design quality and artistic taste.

At this time, the designer presents to us an elegant, unobtrusive, fashion, new living space. Although the project bases on European style, the overall space exudes a unique charm of modern art. Modern visual form language is cleverly added into space, so that users can enjoy the warmth of the sun and blues live in an elegant, beautiful art sense, experiencing their own quiet time.

Through the designer's repeated scrutiny and the careful planning of the space layout, the original functional weakness can be completely eliminated. Simplified modern forms language and traditional, rich linear language form a strong visual contrast. Designer adds an second floor corridor cleverly in the empty space, reasonably divides the vertical space, forming a varied space feeling and functional areas.

Designer boldly uses restrained, unobtrusive light gray as a background color. The clean, jumping white is acting as embellishment, creating a crisp, simple feeling. The overall atmosphere are fashion, comfort and restrained. The relevant design elements emphasize resident's calm, tough, wisdom and profound life experience and artistic taste, showing the space's three-dimensional feeling and unique cultural taste, meanwhile also meets the owner's pursuit of high life quality and comfortable house.

# Lake Enjoyment No 1(I)

观湖壹号（一）

设计单位：上海无相室内设计工程有限公司
设 计 师：王兵、徐洁芳
软装设计师：李欣
项目面积：110 ㎡
摄 影 师：张静

Design Company: Shanghai Wuxiang Interior Design Engineering Co., Ltd.
Designers: Wang Bing, Xu Jiefang
Soft Decoration Designer: Li Xin
Project Area: 110 m²
Photographer: Zhang Jing

设计师一如既往地坚持发掘生活中的美好之处，将本案设计成一个人与人交往的舞台，一个舒适的休息空间，一个有艺术气息的让人难以忘怀的场所。在空间布局中最大限度地满足住户的生活和享乐功能。白色作为主色调贯穿在设计中，运用简洁、明快的设计手法将浪漫的法式线条与中式的木饰面及皮革搭配在一起。金属和玻璃的运用使空间更加灵动。大量的新中式风格的家具，使空间更添古典韵味，彰显出空间的奢华感。

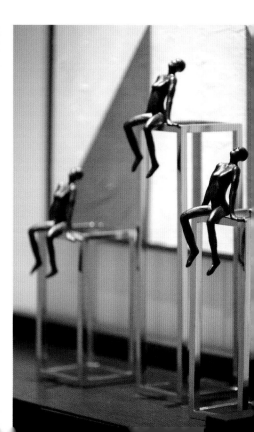

Designers keep to exploring the beauty of life. The case is designed as a stage for human interaction or a comfortable lounge space or an art unforgettable place. In the space layout, designers try to maximize the satisfaction and pleasure of households living functions. White as the main color is throughout the design. Designers use the simple, clean design approach to match the romantic French line with the Chinese style wood finishes and leather. The use of metal and glass makes the space more agile. A large number of new Chinese style furniture add classical charm to the space, highlighting the sense of luxury.

# Mandarin Oriental Apartments

树影下之悠然

设计单位：P+P 室内建筑设计有限公司
设 计 师：廖奕权
项目面积：306.6 ㎡
主要材料：大理石、木、镜

Design Company: PplusP Designers Limited
Designer: Wesley Liu
Project Area: 306.6 m²
Major Materials: Marble, Wood, Mirror

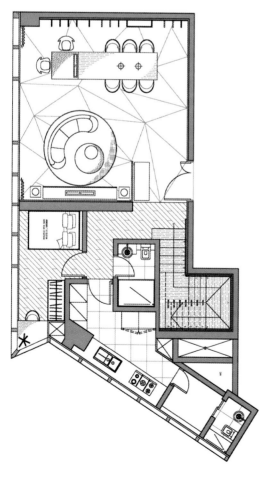
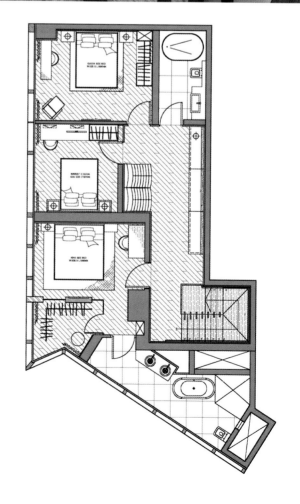

玄关虽不见任何植物和盆景，却处处有自然气息。木质屏风、叶子形状的装饰，拼凑起来似错落枝叶。放眼望去，屏风自入门处起连绵不断，弧形线条至玻璃窗才戛然而止。明亮的空间，辅以顶棚的镜面装饰，质感朴实的茶几，让人仿佛置身于树荫之中，悠然徜徉，感受美不胜收的自然风光，放空身心一扫生活的烦扰。

入门可见的厅堂以暖色调温暖人心，木色与橙黄色调铺饰墙身。客厅中，茶几、沙发、地毯、灯具、屏风，刻意营造出弧形的线条。餐厅以沉稳平实的线条来装饰，用大理石、镜子、实木等材质共同拼凑、互为表里。一切起源于想象，居于内，又形同自然深处，纵使形态、

个性不一，亦在同一环境中相互交融、和谐共存。

转角处，富有艺术感的装饰品无处不在，不着痕迹地提升空间气质，又恰到好处地留白。拐过让人心旷神怡的地下厅堂，来者放轻了脚步，楼梯间，以玫瑰金不锈钢屏风作为过渡。上层的休息区域以浅色木地板营造温润自然感，房间充斥着木色，辅以变化细腻的灯光效果，铺陈睡眠的氛围。

Though the entrance has no plants or bonsai, there isa natural flavor every where. Wooden screens, leaf shape decorations are pieced together like scattered leaves. Looking ahead, the screens are stretching from the entrance, and the curved lines are halted to the glass window. Bright space is complemented with ceiling mirror decoration; the simple textured table makes people feel like being in the shade of trees.People can leisurely wander while enjoy beautiful natural scenery, relaxing the body and throwing away physical and mental annoyance.

Visible entry hall warms people's heart with warm color; wood and orange hue decorate the wall. In the living room, tea table, sofa, carpet, lighting, screens are deliberately creating a curved line. The restaurant is decorated with calm plain lines and it uses marbles, mirrors, woods and other materials. Everything is originated from imagination, inner world and deep nature; even shapes and personalities are different, they also interact in the same environment, coexisting harmoniously.

The corner is full of artistic decorations which enhance the space temperament invisibly, and also leave blank to the point. Walking and turning through the fascinating underground halls, visitor stread lightly.In the stairwell, rose gold stainless steel screen is a transition. Upper lounge area with light wood floors create a warm and natural feeling.The room is filled with wood color, combining with changing delicate lighting effects and sleepy atmosphere.

# Oriental Bund Duplex Unit

泷景复式户型

设计单位：进思设计有限公司
项目面积：365 m²

Design Company: ELEVATION PARTNERS CO. LTD.
Project Area: 365 m²

本案为五房两厅，客厅和餐厅设置在下层，而卧室设置在上层，设计师清晰布局了各个功能空间。除此之外，还有独立电梯直连户型的自主空间，电梯玄关连着别致的景观花园。

本案以摩登优雅为设计主题，为空间营造高雅品位，配合灯光效果使空间拥有华丽、高雅的格调。客厅中，特色水晶吊灯的倒影浮现在意大利云石地台及墙面上方，更显气质不凡。开放式厨房的设计融入到餐厅之中，透光玉石打造的酒柜配衬深色大理石为主调的空间，形成了强烈对比，是餐厅中的一个亮点。装饰材料选用纹理丰富的材质，配以现代欧式家具及灯具，带来了淡雅的异国风情，强化空间的瑰丽感觉。设计师特以知名家具品牌"Armani Casa"的设计概念加入主卧中，展现具有国际视野的业主对高质量生活的要求，彰显出业主的个性、品位。

The case is five-bedroom-two-living room structure, the living room and dining room are set on the lower floor, while the bedroom is set on the upper floor. Designer clearly lays out various functional spaces. In addition, there

are independent elevator directly connects to the separate space. The elevator entrance is attached with the unique landscape garden.

The case's theme is modern and elegant, adding elegant taste to the space. The elegance and lighting effects make the space a gorgeous, elegant style. In the living room, featured crystal chandelier's reflection appears on the Italian marble stages and walls. Thus the space seems extraordinary. Open kitchen design is added into the restaurant. Translucent jade wine cabinet design and dark marble color space forms a strong contrast. It is a stark highlight in the dining room. Decorative material selects rich textured material and modern European furniture and lighting, bringing elegant exotic feeling, highlighting the spectaculars of the space. Designer adds the well-known furniture brand "Armani Casa"'s design concept into the master bedroom, showing the global-visual owner's high requirements of life, highlighting the owner's personality and taste.

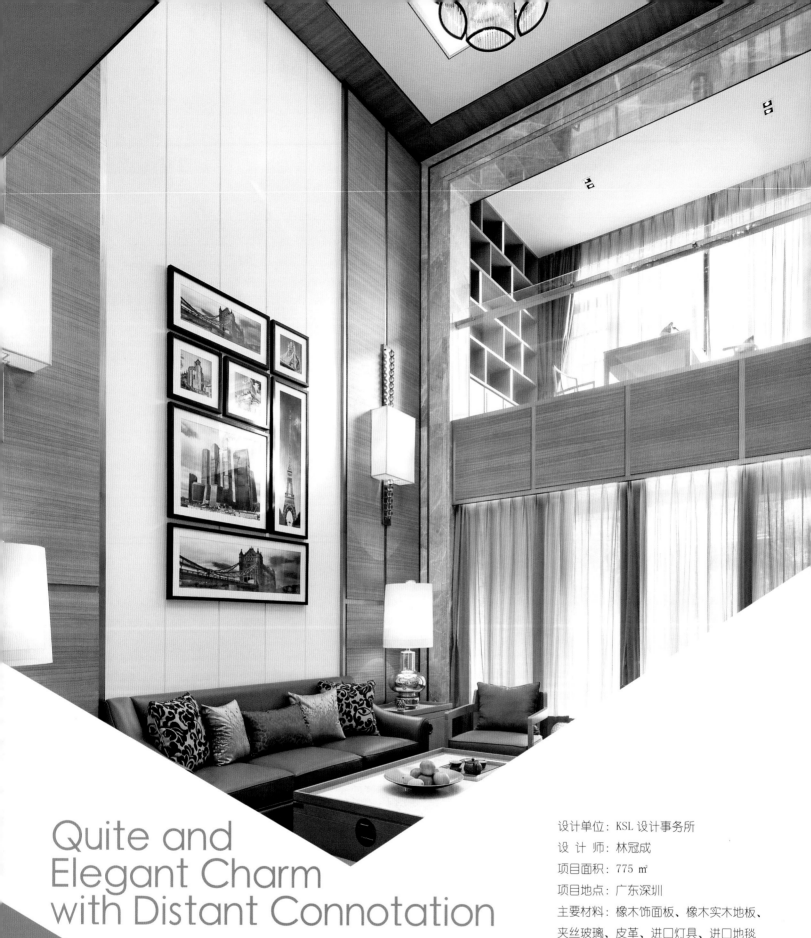

# Quite and Elegant Charm with Distant Connotation

清雅灵韵 内敛悠长

设计单位：KSL 设计事务所
设 计 师：林冠成
项目面积：775 m²
项目地点：广东深圳
主要材料：橡木饰面板、橡木实木地板、夹丝玻璃、皮革、进口灯具、进口地毯

Design Company: KSL Design Firm
Designer: Andy Lam
Project Area: 775 m²
Project Location: Shenzhen, Guangdong
Major Materials: Oak veneer, Oak wood floors, Wire glass, Leather, Imported lamps, Imported rugs

大象无形,不饰雕琢,灵韵悠长。此别墅样板房设计去除了繁复的藻饰,采用雅白与橡木饰面的顶棚搭配大理石的色泽纹饰,清雅、自然的韵味内蕴深长。挑高的空间设计及灵活巧妙的布局,塑造出大宅的稳健气度。极富质感的进口陈设架构出空间的精致度与人文气质,焕发出多重文化的意蕴神采,晕染出深隽、内敛的现代简约空间。

客厅中挑高的空间设计顿显明亮、大气,在柔和灯光的调和下,精美、奢华的进口灯具与皮革等高档材质相互辉映。书房以木色为主、灰色为辅的基调营造了古朴、安静的书香氛围,清新的蓝色给予视觉另一番惊喜,简洁而有力度的设计提升了空间的格调。

三层主卧独特的挑高空间及稳定的三角顶棚造型,巧妙地带来了截然不同的视觉享受。整面的落地窗营造出通透、敞亮的生活空间,光影随时间流动更显静谧。

二层卧室用橡木实木地板打造出简单、舒适的休憩空间,湖蓝石墩

和白色陶瓷在灯光下越发剔透，整个空间看似不经意的搭配，带来了清醇而又悠远的感受。另一间卧室中线条简练的实木家具和精致的陈设在灯光烘托下更显质感，无阻碍的窗景从落地窗向室内延伸，让业主的身心享受到微风轻拂的自在与惬意。浴室以简洁流畅的设计线条勾勒出整洁、轻松的环境，让一天的繁忙得到舒缓。

红酒雪茄吧中博古架和奢华皮革家具带来文化的多重韵味，璀璨剔透的水晶灯让古典韵味与现代简约风格巧妙融合。

It is said that the most beautiful thing is invisible, without decorative sculpture. It uses elegant white color and oak covers to match the marble color and texture. Fresh and natural charm lasts for a long time. High space design and clever, flexible layout create a steady mansion. Highly textured import display rack creates the space's elegance and human temperament, waking up multicultural meaning, forming deep, restrained modern simple space.

High space design in the living room makes it bright. Under the soft light, exquisite, luxury import lamps and leather echoes with each other. Study mainly bases on wood color, and color gray is used as ornaments, they create a simple, quiet, sophisticated atmosphere. Fresh blue color gives another surprise on vision, concise design enhances the space's style.

Unique high space in the main bedroom on the third floor and stable triangle ceiling subtly brings different visual enjoyment. The entire french window creates a transparent, light and spacious living space. Light quietly flows over time.

Second floor's bedroom uses oak wood floor to create a simple,

comfortable open space. Blue rock and white ceramic seems lighter and clearer under the light. The entire space seems casually combined, bringing mellow and distant feel. In another bedroom, the simple lines fine wood furniture and delicate furnishings seems beautiful under the light. Unobstructed window extends from french windows to the indoor, so that the owner can enjoy the breezy comfortable and cozy. Bathroom has concise and fluent design lines sketching out clean, relaxing environment to ease the busy day's pressure.

Wine and cigar bar's shelf and luxurious leather furniture bring multiple cultural charm, bright crystal light combines classical charm and modern simplicity.

# House in Shatin

沙田独立屋

设计单位：Millimeter Interior Design LTD.
设 计 师：廖子扬
项目面积：357 ㎡
项目地点：香港
主要材料：水泥、铝塑板、铁器

Design Company: Millimeter Interior Design LTD.
Designer: Michael Liu
Project Area: 357 m²
Project Location: Hong Kong
Major Materials: Cement, Aluminum plate, Iron

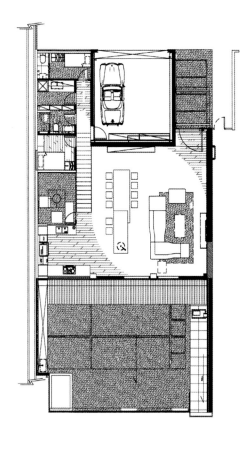

能将一所旧式别墅变为现代都市风格,还不留痕迹地保留其原有结构,的确是一件值得称道的作品。本案采用一种绿色可持续发展的方式在原址上进行大胆改造,将占地 167 ㎡、屋龄 40 年之久的一所旧别墅变为一套 357 ㎡ 的舒适又宽敞的现代住宅。

为了减少对环境的影响,设计师用节能铝塑板将别墅外墙重新构建于旧有风格之上,如此一来原有结构也得以保全。这不但降低了重建成本、避免了浪费,还可以将热量阻隔在别墅之外,从而减少了空调的使用率。

设计师成功地将这个两层别墅改为具有一个车库、一间起居室、一个餐厅、一座花园、两间客房、两间客房卫生间、两个辅助套间、一个带有宽敞衣橱的主卧套间及一间书房的住宅。设计打破了室内外分界,通过巨大的窗户和别墅中间的透明屋顶极好地采用了自然光。另外,在别墅后方延伸出一块宽阔的场地,完美地将一块户外空间融入其中。

The case's uniqueness lies in changing the old-style villa into a modern urban style, and retaining their original structure. In this case, designer adopts a way of green and sustainable development, boldly transforms the 167 $m^2$, 40 years old villa into a 357 $m^2$ comfortable and spacious modern residence.

To reduce the impact on the environment, designer uses energy-saving aluminum plate to re-built villa's facade on the old style. Therefore the original structure has been preserved. This not only reduces the cost of rebuilding, avoids waste, but also separate the heat barrier outside the villa, thus reducing air conditioning usage.

Designer successfully transforms the two-tier villa into a house with a garage, a living room, a dining room, a garden, two guests rooms, two bathrooms, two secondary suites, a master suite with a spacious wardrobe and a study. Design breaks the boundary between inside and outside, through huge windows and transparent roof makes use of the natural light. In addition, designer extends a wide space behind the villa, perfectly adds a piece of outdoor space into the entire design.

# Mangrove West Coast

红树西岸花园

设计单位：于强室内设计师事务所
设 计 师：于强
项目面积：403 ㎡
项目地点：广东深圳
主要材料：月影木饰面、白色亚光漆、纯白人造石、黑砂钢

Design Company: Yuqiang & Partners
Designer: Yuqiang
Project Area: 403 m²
Project Location: Shenzhen, Guangdong
Major Materials: Moony wood finishes, White matt paint, White artificial stone, Black sand steel

本案为拥有一线海景的复式滨海大宅，有着无可比拟的景观优势。在平面布局过程中，如何令格局开敞、通透，把最具优势的景色引入室内，成为非常重要的设计因素。

一层的公共区域，客厅与餐厅间的开敞格局，使两个相对独立的空间形成互动，将景色更多地引入，在无形中扩大了空间感。主卧及相对私密的休闲区设在二层，对比一层的"动"，二层突显了更多的轻松与安静。

在材质的搭配方面，运用了大量的木、丝麻及部分有着漂亮肌理的石材，营造出轻松、休闲又不失亲切的滨海大宅氛围。

一层平面图

This case is a duplex house with coastal line sea view. It has an unparalleled landscape advantage. In the layout process, how to make the pattern open, transparent, and leads the most advantage view into the room is a very important design factor.

One the first floor public area, open pattern of living room and dining room make the two relatively independent space interacting. The scenery virtually expands the space. The master bedroom and relative privacy recreation area is located on the second floor, comparing with the first floor's "action", the second floor highlights more relaxation and quietness.

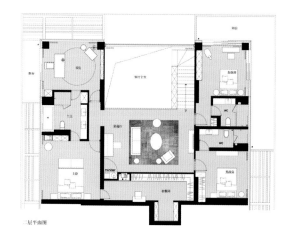

二层平面图

In terms of material, designer uses a lot of wood, linen and partial beautiful texture stone, creating a relaxed, casual yet intimate seaside mansion.

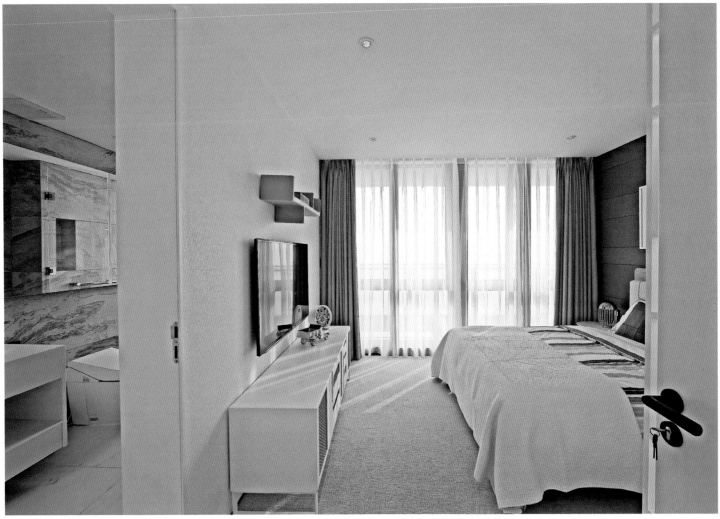

# North Point Breamar Hill Mansion

北角赛西湖

设计单位：COMODO Interior & Furniture Design CO.,Ltd.
设 计 师：王智衡
项目面积：117 ㎡
项目地点：香港
主要材料：木纹饰面、浅灰色壁纸

Design Company: COMODO Interior & Furniture Design CO.,Ltd.
Designer: Alain Wong
Project Area: 117 ㎡
Project Location: Hong Kong
Major Materials: Wood finishes, Light gray wallpaper

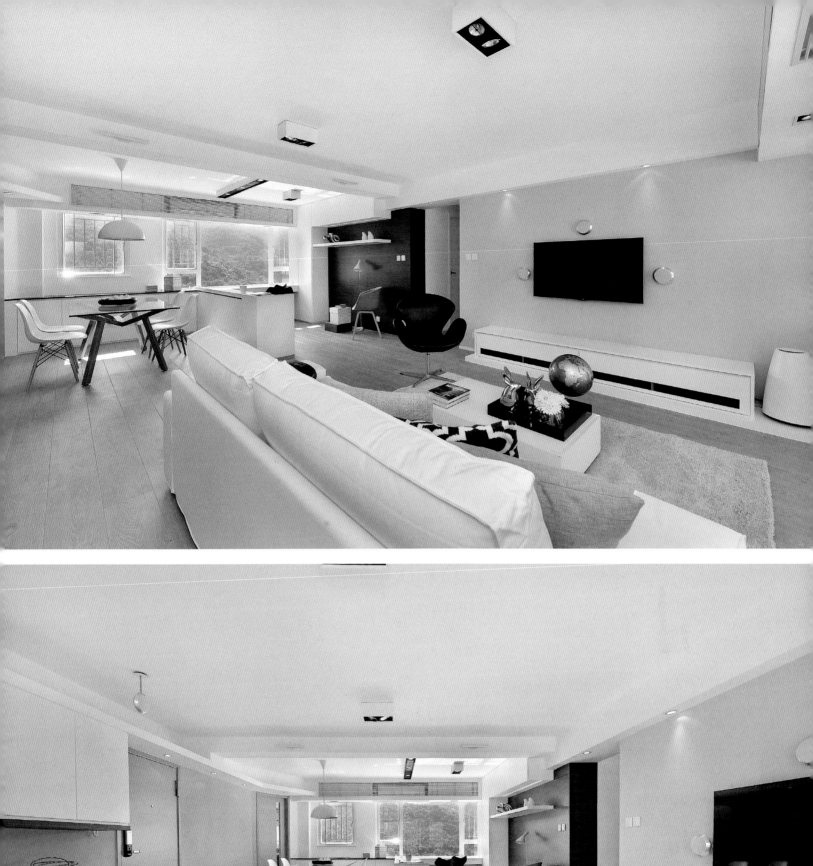

设计师为整个空间选用了浅灰色壁纸，再配合巧妙的灯光设计就像为屋内洒上一片淡淡的阳光，令居住空间更显温暖舒适。在客、餐厅的中间有一道把其一分为二的横梁，设计师分别在它的两边安装了两盏壁灯，让光线均匀投射在横梁上，保持了空间的完整性。

在原有的室内格局上，餐厅因为和厨房共享了部分的面积而显得比较狭窄，和另一边的客厅面积对比明显，减少了空间的美感。设计师把客、餐厅和厨房的面积重新规划，把厨房推后，拆除原有的墙身，令客厅、餐厅空间变得平整，让更多的自然光线可以照射到室内。在客厅、餐厅的另一边则是用书柜分隔出一个开放式的工作间，后方是一面较深色的木纹饰面，用以陪衬稍后搬入的钢琴。

设计师大致上没有改动主人套房的格局，只是把门向前移了少许，方便在后方放置衣柜，增加储物空间。房间的另一边则是业主的工作空间。为了减少对在床上休息的家人的影响，设计师特意设置了一道屏风，以便把电视分隔开来。

在客卧设计中，设计师把两间客房中间的墙拆去，用两个衣柜作部分分隔，更加灵活地运用了两个房间。客房吊柜的底部安装了灯槽，让中间的部分更显明亮。床头的位置则铺上米色扪皮，平衡房间的色调。设计师更在双层床的一角安装了阅读灯，这样就不会影响到家人休息。

Designer selects light gray wallpaper for the whole space, coupled with clever lighting design which is like sprinkling sunshine into the house, so that the living space is warmer and more comfortable. In the middle of the living and dinning room, there is a crossbeam separating them apart. Designer installs two wall lamps at both sides of the space, so that the light is homogeneously shinning on the crossbeam, maintaining the integrity of the space.

In the original interior structure, restaurant and kitchen seem relatively narrow since they share partial area, and contrasting with the living room area on the other side, the aesthetic is reduced. Designer re-plans the living room, dining room and kitchen, and pushes the kitchen back, removing original wall, making the space to the living room and dinning room flat, letting more natural light into the interior. On the other side of the two rooms there is an open workplace separated by a bookcase, the rear is a dark color finishing aiming to foil the piano.

Designer generally has not change the pattern of the master's suite, except moving the door to forward a little bit, in order to put the wardrobe in the rear and increase the storage space. The other side of the room is the work space of the owner. In order to reduce the impact on the family resting in bed, the designer deliberately sets up a screen to separate the TV.

In the guest bedroom design, the designer moves away the middle wall between the two guest rooms, using the two wardrobes as partition and the two rooms are used more flexibly. On the bottom of the guest rooms, there installs lamp groove, so that the middle part is more bright. The location is covered with beige leather, balancing the tone of the room. Designer installs reading lights in the corner of a bunk bed, so it will not affect the family rest.

# Pok Fulin Residence Bel-air

薄扶林贝沙湾

设计单位：COMODO Interior & Furniture Design CO., Ltd.
设 计 师：王智衡
项目面积：167 ㎡
项目地点：香港
主要材料：木纹地砖、壁纸

Design Company: COMODO Interior & Furniture Design CO.,Ltd.
Designer: Alain Wong
Project Area: 167 m²
Project Location: Hong Kong
Major Materials: Wood floor tiles, Wallpaper

室内空间用大地色系作为主色调，运用浅色调搭配浅啡色木纹地砖，营造出温馨的气氛。追寻简单、舒适的业主和设计师一拍即合，双方都认为居住空间应该简单一点，以便调剂日常急促的生活节奏。一踏进这个四口之家的温馨空间，就感觉到一阵缓缓流动的温暖空气，让身处其中的人可以放下烦恼，悠闲地与家人共享美好时光。

客厅里选用了不同材料去加强质感，设计师在毛绒地毯上摆放一张线条简单的茶几，在质感上制造一个鲜明对比。另外为了让摆放钢琴的地方不显得突兀，特意把墙刷成较深的颜色，将乐器融入客厅当中。设计师把连接女儿房和客厅的墙壁向客厅方向移出些许，平衡另一边的柜子突出的部分。客厅、餐厅只用了一盏吊灯作装饰，其余都是用顶棚灯槽和射灯均匀投放光线，令室内更显温暖。

设计师为偶尔要晚归的业主在走廊的位置安装了三盏地灯，方便其出入房间。主卧用大衣柜分隔出衣帽间和书房。和客厅一样，主卧也尽享了一望无际的海景，设计师除了把睡床摆放成面向海洋的方向外，更在窗边放置了一张紫红色的沙发椅，使之成为一个独立的阅读空间。另外，设计师还为喜欢观赏电影的业主预留了放置电视和音响的位置。设计师在梳妆台的后方设置了一道屏风，增加了装饰效果的同时，让女主人有较私密的装扮空间。

至于两位小朋友的空间则注入更多的色彩配置，让简洁的空间散发点点童真。儿子的房间布置了一面特色墙，贴上淡蓝色的星星壁纸。女儿的房间就加入粉红色的元素，窗边摆放了一张桃红色、线条圆润的椅子。低调的色彩配置可以随着不断成长的小朋友的喜好而改变，只需一些简单的配件改动就可以改变房间的整体气氛，足以应付小朋友成长期的快速改变。另外考虑到小朋友喜爱画画，设计师在两间房都装上白板，让他们可以尽情发挥创意，画出他们脑中的小宇宙。

Interior space uses earth tones as the main color and uses light colors with light brown wood slabs, creating a warm atmosphere. Designer and owner both pursue simple, comfortable style. They believe that the living space should be simpler to adjust the daily rapid pace of life. Stepping into this family of four, one can feel a slowly flowing warm air, in which people can put down troubles and enjoy a leisure family time together.

The living room uses different materials to strengthen the texture. On the plush carpet, designer puts on a simple line coffee table to create a sharp contrast in texture. In addition, to make the piano place unobtrusive, designer deliberately brushes the wall into dark color, adding the instrument into the living room. Designer moves the wall connecting daughters room and living room a little towards living room to balance the cabinet's prominent part on the other side. Living room and restaurant only uses one chandelier for decoration, the rest are ceiling lights and spotlights, so that the room is warmer.

Designer installs three lights at the corridor for the owner. It is convenient for entrance. Master bedroom uses wardrobe to separate the cloakroom and study. In the living room and the main bedroom, people can enjoy the vast sea view. Designer puts the beds towards the sea, in addition, places a purple red sofa chair beside the window, making an independent reading room. Besides, designer also set aside a location to place the TV and stereo for the owner. It is

like a small private theater. Designer designs a screen at the rear of the dresser, so that the hostess has more intimate space.

As for the two kids' rooms, the spaces are injected with more colorful schemes. So that the simple space will exude some innocence. Son room has a feature wall pasting with light blue stars wallpaper. Daughter's room has been added pink elements, the window side has a pink, round chair. Unobtrusive color can change with the growth of kids, just a few simple parts can change the overall atmosphere of the room, which is enough to cope with the rapid change of children. In addition, taking into account that kids love drawing, designer installs white board in the two rooms, so that they can develop their creativity, drawing the small universe in their brain.

# Deep in Nature
自然意象

设计单位：P+P 室内建筑设计有限公司
设 计 师：廖奕权
项目面积：244 ㎡

Design Company: PplusP Designers Limited
Designer: Wesley Liu
Project Area: 244 m²

当创意融入生活，碰撞出的火花能够点燃生活中许多有趣的情怀，所呈现的世界更是处处充满惊喜。本案的设计，在求新、求变中加入富有生活趣味的小元素，大玩创意的同时，也兼顾了居住空间的要点，既有欧美居室的创造性思维，又有东方住宅的细腻用心，彰显出自身的与众不同。

富有创意的细节是本案的最大亮点。个性十足的客厅，将撞色与拼接尽情玩味，灵活多变的线条和几何图形，赋予空间艺术生命。餐厅中，空间顶部的抽象树枝张扬、醒目，与纯朴自然的木质桌椅、桌面上几何玻璃内的小绿植相映成趣，将大自然的无限生机带入室内。从顶棚上垂吊而下的吊灯，犹如在空中定格的水珠，构成了一幅自然唯美的画面。休闲阳台用纯天然的材质来打造，清凉的藤椅与原木的个性装饰、实木的地板相处融洽，营造出宁静、自然的氛围。卧室的设计，将各种独特的元素放入室内，玩转创意。贯穿于空间中的装饰画和创意十足的各种小饰物，以及造型别致的置物架，拼凑出许多独特的画面，并为业主带来浓郁的生活气息。

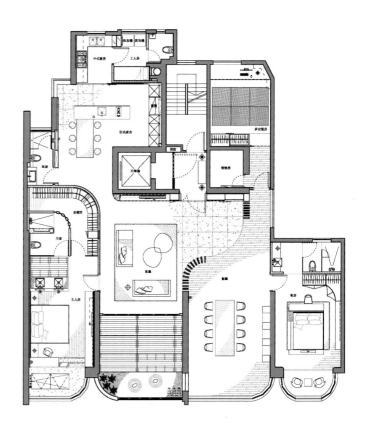

When creativity comes into daily life, the collision of spark can ignite many interesting feelings in life. The presented world is full of surprise. The case design adds small interesting elements into novelty and change. The design is creative and practical. It also takes into account the point of living space. It has European and American creative thinking and also delicate oriental residential intentions, highlighting its distinctiveness.

Creative detail is the biggest highlight of the case. Individual living room contains different colors. Flexible lines and geometric shapes give space artistic life. In the

restaurant, the abstract branch at the top of the space is eye-catching, simple and natural, forming interesting structure with the wood tables and chairs, Green plants in the glass on the table brings natural vitality into room. The chandelier hangs from the ceiling is like water drops freezing in the air, constituting a natural beautiful picture. Leisure balcony is made of pure natural materials. Cool wicker chair and wood personal decoration, solid wood flooring creates a tranquil, natural setting. Bedroom design puts variety of unique elements into the interior. Various paintings throughout the space and small decorations and unique racks are pieced together many unique pictures, bringing strong flavor of life to the owner.

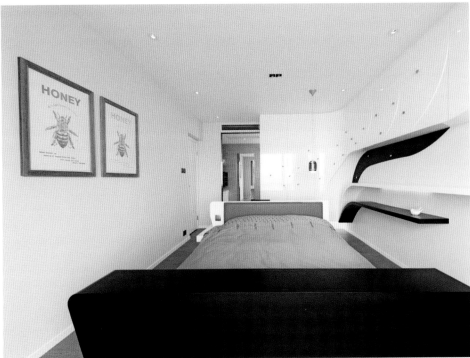

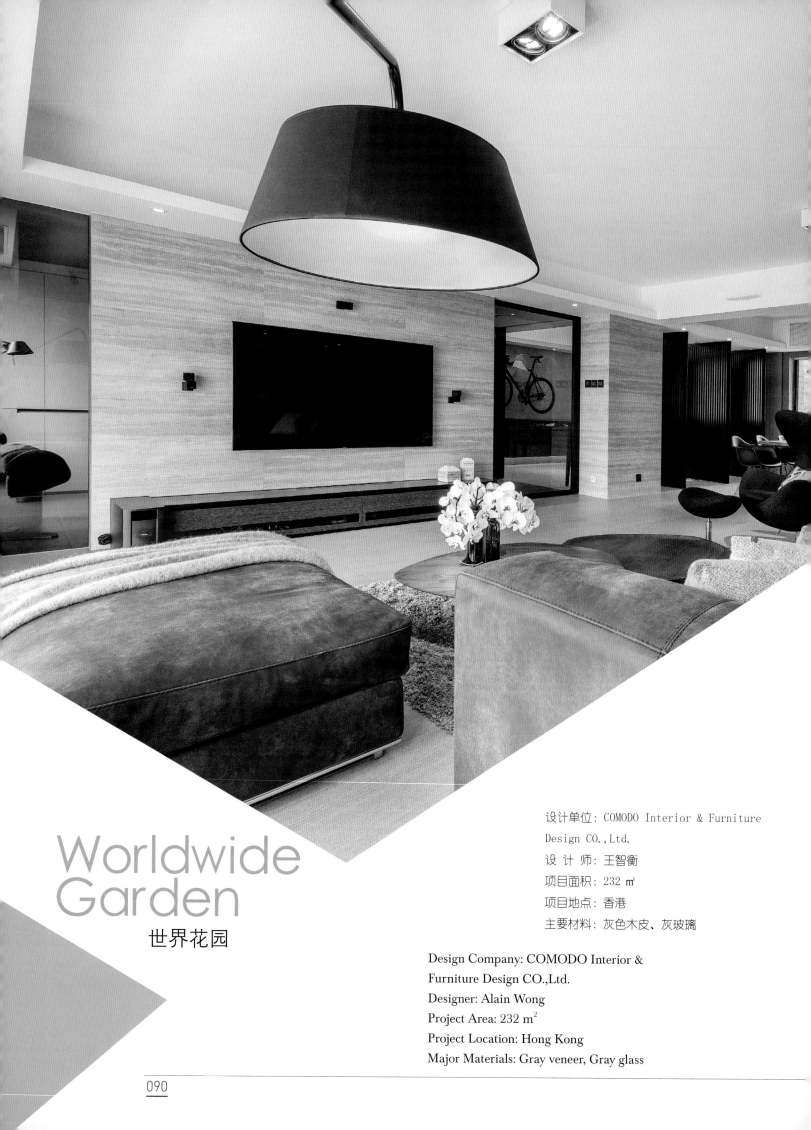

# Worldwide Garden
## 世界花园

设计单位：COMODO Interior & Furniture Design CO.,Ltd.
设 计 师：王智衡
项目面积：232 ㎡
项目地点：香港
主要材料：灰色木皮、灰玻璃

Design Company: COMODO Interior & Furniture Design CO.,Ltd.
Designer: Alain Wong
Project Area: 232 m²
Project Location: Hong Kong
Major Materials: Gray veneer, Gray glass

本案有近三十年的楼龄，上一家的装潢一直保留至今。设计师将用现代的品位去重新设计这个家，以欢迎它新的主人。业主是一个单车爱好者，在设计初期已提出希望在设计上添加单车的元素。另外，他也希望将黑色等较深的颜色定为设计主色调。

考虑到只有业主二人居住，设计师几乎改造了整个空间。在布局上分为两个部分，以客餐厅、书房、娱乐室和厨房组成的公共区域，以及卧房和衣帽间组成的私密区域。整个空间采用业主偏好的灰、黑色调，颜色由客厅到卧房慢慢转深，明确地把私人空间区分了出来。为了善用空间，设计师把原本占了不少位置的走廊的两边墙身移除，把走廊的空间也划为书房的一部分。走廊的位置原本有一道横梁，设计师用假顶棚填平通道的部分，让空间看起来更完整。

设计师希望属于公共区域的三个房间在功能分明的同时又互相紧扣。他先拆掉玄关的屏风，让客厅变得更加开阔。公共空间的隔断设计不能只用玩味去形容，设计师把原本普通的一道门变成可营造不同感觉的艺术装置。最特别的一道是受注重健康、热爱运动的业主的启发，在连接娱乐室和餐厅的隔断上改用三幅犹如车轮一样的屏风，可让业主随着需要转到不同的方向，还可以隐约看到安装在娱乐室的墙壁上的战车。书房连接客厅和娱乐室的部分则用上灰玻璃拉门，让空间更

显通透。当把隔断全都拉开的时候,这几个空间就连接在一起。

推开书房深灰色木皮的暗门,进入隐藏在空间最里面的私人空间,第一眼看到的是一张深啡色的真皮躺椅,让业主可以在睡前看看新出版的杂志。同时为避免横梁把空间分开,设计师干脆在靠浴室和衣帽间的一边都装上假顶棚,把横梁隐藏在其中。

在休息空间的两边分别是浴室和能够媲美高级男装店的衣帽间,两个房间和客、餐厅一样用灰玻璃相隔,让空间与空间更加连贯。虽然猛然一看浴室好像一目了然,但设计师其实把坐厕和浴缸都安排了在梁柱的后面,因此浴室外的绿意反而成为最亮眼之处。

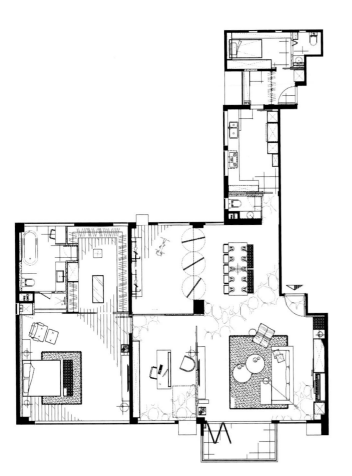

The building exists nearly 30 years, the original decoration has been preserved to the present. Designer will use modern taste to redesign the house to welcome its new owner. Owner is a cycling enthusiast, at the beginning of the design, he wished to add a bicycle element on the design. In addition, he also wants to use dark color as the main color of the design.

Considering that there are only two members in the house, designer almost transforms the entire space. The layout is divided into two parts, one is composed by dining room, study, recreation room and kitchen, the other is composed by bedroom and cloakroom. The entire

space uses the owners' favor gray, black tone. Color is turning darker from living room to the bedroom, clearly highlights the private space. In order to make use of the space, designer removes the corridor walls at both sides, the corridor space has been divided as part of the study. Corridor's position originally had a beam, now designer fills the tunnel with false ceiling, so that the space seems more complete.

Designer hopes the three rooms in the public area are interlocked while have clear functions. He first removes the porch screen, makes the living room more open. Partition design in the public is not only interesting. What is more, the designer makes the original door into an art installation that can create different feelings. The most special partition is inspired by the owner who interested in health and sports. The partition links the entertainment room and dinning room are three pieces of wheel-like screens which allow owner turning to different direction, and one can vaguely see tanks installed on the wall in the recreation room. The link part of living room and entertainment room is using gray glass sliding doors, so the space is more transparent. When the partition have been open, the above spaces can be connected together.

Open the study's dark gray wood cover door, one can enter into the hiding space in the private space. First thing one can see is a dark brown leather chair, for the owners' magazine time before go to bed. At the same time, in order to avoid to separate the space with the beam, designer installs false ceiling on both the bathroom and cloakroom, the beam is hidden inside the two rooms.

On the two side of the resting place are bathroom and cloakroom. The two rooms are divided by gray glass just like the living room and dining room, the spaces are linked together. Although the bathroom is clear at a glance, in fact designer arranges toilet and bathtub behind the beam, so the outside green has become the most attractive place.

# Dynasty Heights II
## 帝景台（二）

设计单位：ANSON CHENG INTERIOR DESIGN LTD.
设 计 师：郑秋基
项目面积：223 ㎡
项目地点：香港
主要材料：木纹地砖、地毯、人造石

Design Company: ANSON CHENG INTERIOR DESIGN LTD.
Designer: Anson Cheng
Project Area: 223 m²
Project Location: Hong Kong
Major Materials: Wood floor tiles, Carpets, Artificial stone

设计讲求美学，但若出发点仅为美学，忽略了业主的需求，未免本末倒置。

本案业主为一对退休夫妇，在这个居所已住十余年了。初步规划后，不打算再对户型做大改动，所以这次设计的重心在于顾及他们的生活模式及习惯，塑造易于打理、舒适、宽敞的环境为先。设计应着重实用性，在潮流的元素与舒适度之间巧取平衡，充分利用空间，缔造家的感觉。

考虑到居所只有二人居住，部分多出的房间并未得以充分利用，使得原来双厨房的格局很浪费。设计师重新调整了间隔，房间的数量由原来的四房减到两房，主人房则以三个房间打通而成，空间宽敞。厨房作半开放式规划，仅以玻璃趟门划分，极具通透感，并且与用餐区域、户外花园相连，无限惬意。

下层的厅堂、厨房，甚至客厅皆以仿木纹地砖铺饰，除了在视觉上会感觉柔和、舒适，增强统一感之外，也相对容易打理。至于在色调运用上，"简约"的设计主题多以素白色调为主，设计师则多选择自然色系，如以奶茶色手扫漆，缔造温暖感萦绕的室内环境。环视客厅，清简的陈设绝不单调，特色墙以双色木板勾勒出线条，卡其色饰面木板从特色墙延伸，经过卫生间、厨房，借着线条勾勒墙身，巧妙地为简约的厅堂布局增添不凡层次。

Design emphasizes on aesthetics, but if the start point is only aesthetics, ignoring the owner's needs, the results may be upside down.

The case's owner is a retired couple, who have lived here for over ten years. After preliminary planning, we do not intend to change the apartment a lot, so the design's focus lies in their lifestyle and habits. The environment should first be easy to live, comfortable and wide. Design should focus on practice, making a balance between fashion elements and comfort. The designer makes full use of the space to create a homey feeling.

Taking into account that the home only contains two people, some rooms have not been fully utilized. Therefore the original double kitchen pattern is a waste. Designer restructures the interval, the number of rooms have reduced from four bedrooms to two bedrooms. The master's bedroom has been composed by three rooms. The space is wide. The kitchen is half-open, using glass door as partition.

The space is full of transparency, and it is connected with dinning area and outside garden.

The lower floor hall, kitchen even the living room has been covered with wood texture tiles. In addition to the soft, comfortable vision which enhances the sense of unity. It is relatively easy to clean. As for the use of tone, "simple" design theme is based mostly on plain white tone. Designer often uses natural color, such as milk-tea color brush paint, to create a warm indoor environment. Looking around the living room, simple display is not monotonous. The feature wall uses double-color to sketch out the line; khaki finishes wood plate extends from feature wall through the bathroom and the kitchen. The line sketches the wall, and it cleverly adds extraordinary hall level to the layout.

# Lake Enjoyment No I(II)

观湖壹号（二）

设计单位：上海无相室内设计工程有限公司
设 计 师：王兵、徐洁芳
软装设计师：李欣
项目面积：90 ㎡
摄 影 师：张静

Design Company: Shanghai Wuxiang Interior Design Engineering Co., Ltd.
Designers: Wang Bing, Xu Jiefang
Soft Decoration Designer: Li Xin
Project Area: 90 m$^2$
Photographer: Zhang Jing

本案是一个 90 ㎡ 的公寓，为三房二厅的布局。空间以灰色为基调，用白色和深灰色进行搭配，厨房的茶色玻璃移门巧妙地显示出空间的开合和相互关联，并留出空间以放置饰品。客、餐厅每件家具的选择都十分用心，整体空间显得宽大而且温馨。卧室的衣帽间也沿用了这个方法，空间非常实用。本案的亮点是有一个小巧、精致的多功能空间，可以作为书房、客卧使用，还可以作为茶室使用，业主甚至可在此冥想、练习瑜伽。

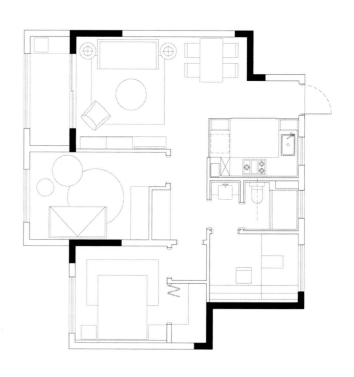

This case is a 90 m$^2$ apartment of three-bedrooms-two-living rooms' layout. Space uses gray as tone, matched with white and dark gray. The kitchen's tinted glass sliding door subtly shows the opening and closing of the space; besides, the designers leave full space to display decorations. Each choice in the living and dinning room are full of the designers' heart. The whole space seems large and warm. Bedroom's cloakroom continues to use this method, making the space very practical. The highlight of this case is a compact, elegant, functional space, which can be used as study room, guest bedroom or tea room; the owner can even has meditation and yoga in this room.

# Hannover Court
## 恒懋大楼

设计单位：擎风设计
设 计 师：林子康
项目面积：185.8 ㎡
项目地点：香港
主要材料：定制木器家具、复合地板、壁纸

Design Company: Moderne Design
Designer: MAX LAM
Project Area: 185.8 m²
Project Location: Hong Kong
Major Materials: Custom wood furniture, Flooring, Wallpaper

原本是四房两厅的格局,设计师将之摇身一变,变成了三房两厅的和谐爱巢。从门口进入,就能看见开放式的厨房、客厅连接露台的景观,开敞而气派。原本是客房的位置改造为餐厅,利用了主力柱位,建了渗光地台,打造了一个VIP角落。而书房用了玻璃间隔,让阳光透过书房射入走廊,减少长长走廊的压迫感,充分地利用了日光。通透的间隔及空间的合理运用,使客厅、餐厅、书房及厨房,虽自各自分隔但又紧密连接。业主夫妇即使在家中不同角落工作、烹调、看电视,都好像守在一起。室内色调以白色、木色及蓝色为主,米色沙发搭配浅木色地板,衬以白色饰物柜,用蓝色地毯作点缀与餐厅的蓝色墙身呼应,感觉舒适、和谐。

The house originally has four bedrooms and two living rooms. Designer transforms it into a three bedrooms and two living rooms' harmony love nest. Entering from the door, one can see the open kitchen, the wide and open linked living room and terrace. The original guest room now is the dining room. And the main column position and infiltration lamp floor now is a VIP corner. The study uses glass partition, the sunshine enters the corridor, reducing the pressure of long corridor, and make full use of daylight. Rational use of transparent partition and space makes living room, dining room, study and kitchen, separately but closely linked together. The owner couple is working at different corners of the house. They cook food and watch TV together. Interior tone is based on white, wood and blue. The match of beige sofa and light wood floors, white cabinets and blue carpets and walls make the space comfortable and harmonious.

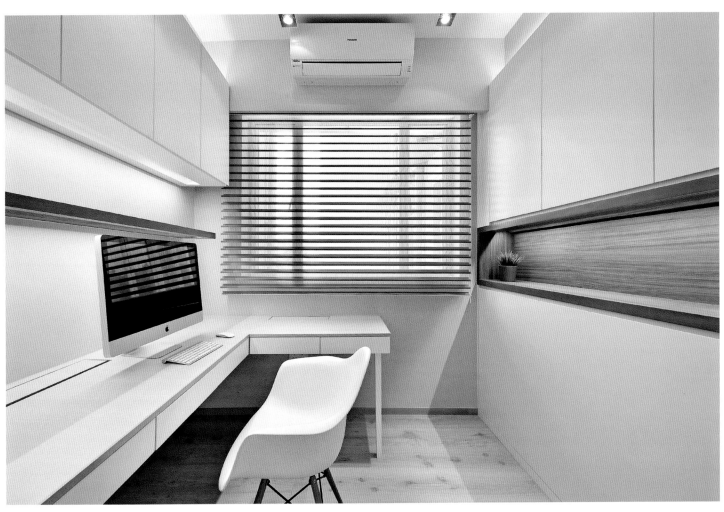

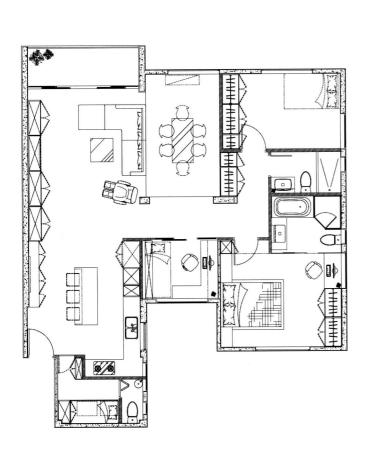

# The Westminster Terrace
皇璧

设计单位：Danny Cheng Interiors Ltd.
设 计 师：郑炳坤
项目面积：356 ㎡
项目地点：香港

Design Company: Danny Cheng Interiors Ltd.
Designer: Danny Cheng
Project Area: 356 m²
Project Location: Hong Kong

这是一个复式住宅，设计师以精致的设计突显出这里非同一般的感觉。

通过一道电动大门，带你走进这个富有气派的居所，大门旁的镜面墙身在增加空间感之余，也反映了餐厅的景象，增添了些许趣味。开放式的厨房配以设有洗手盆的长形白色餐桌，带给人开放互动的感觉。木皮旋转门成为客厅的焦点，可将客厅划分出一个隐私度较高的空间给业主或客人。镜钢电视柜在供业主摆放东西之余，也提升了客厅的格调，突显楼底高的优势。设计师以大理石作为客厅、餐厅的地台，并延伸至户外露台，营造出高雅、时尚的氛围。客厅的花形图案地毯配合露台的植物、流水声，增添了大自然的气息与写意的气氛。从厨房、餐厅到客厅，没有间隔墙身，整个空间富有通透感及连贯性。

楼梯及卧室铺设了木地板，为休息的空间营造温暖感。主人房以棕色为主色调，营造出和谐的氛围。床背以扪布作墙身，增添了舒适的感觉。偏厅提供了充足的空间给业主休息及放松心情。花形图案地毯为主人房增添生气，亦与客厅互相呼应。以木皮为主要装饰材料的衣帽间，充分地利用空间，让业主可摆放大量衣物，配合玻璃面的独立柜，可陈列首饰，手表等饰物，美观与实用性皆备。

This is a duplex apartment. Designer uses exquisite design to highlight the general feeling here.

Together with electric gates take you into this stylish home, large door mirror walls, apart from increasing the sense of space, but also reflects

the vision of the restaurant, adds a little fun. The open kitchen has a wash basin with elongated white table, give people the feeling of open and interactive. Veneer revolving door has become the focus of the living room, the living room can be divided into a higher degree of privacy to the owner or guest space. Mirror Steel TV cabinet to allow owners to put things in, but also an enhanced living room style, highlight the ceiling height advantage. Designer marble as the guest, the restaurant floor, and extends to the outdoor terrace, creating an elegant, stylish atmosphere. The living room carpet with flower-shaped pattern of plants, the sound of water terraces, adding natural flavor and decent atmosphere. From the kitchen, dining room to the living room, no space wall, the entire space full of sense of vision and consistency.

Stairs and bedrooms are covered with wood floors, creating a warm sense. Master bedroom is covered with brown tone, creating a harmonious atmosphere. Bed back uses cloth as the wall, adding a comfortable sense. Side living room provides ample space to the owners to rest and relax. Flower-shaped pattern carpet adds vitality to the master room, echoing with the living room. Wood leather cloakroom is mainly using wood cover, the large space can contain lots of cloth. The glass surface independent cabinet can display jewelry, watch and other accessories with beauty and practicality.

# Hong Lok Yuen
## 康乐园

设计单位：ANSON CHENG INTERIOR DESIGN LTD.
设 计 师：郑秋基
项目面积：297 ㎡
项目地点：香港
主要材料：地毯、人造石

Design Company: ANSON CHENG INTERIOR DESIGN LTD.
Designer: Anson Cheng
Project Area: 297 m²
Project Location: Hong Kong
Major Materials: Carpet, Artificial stone

餐厅跟入门区并排分布，只有一墙之隔。多余的庭园间隔墙拆掉了，苍翠身影因而映入室内。设计师打造了开放式的厨房，使视觉开阔。电视背景墙设计得较为低调，仅以方形层板修饰，却成功地衬托出吊灯、餐桌及餐椅。事实上，即使餐桌和吊灯以实木或幼木条制造，在一片黑、白、灰色中仍可突显出来。

厨房和餐厅组成了一个"吕"字形空间，餐桌、工作台和橱柜平行排列，维持走道畅通，墙身与柜架以白色居多，营造出空间感和明亮度。餐椅、吧台和吧椅同选黑色，发挥点到即止的陪衬作用。厨房分为两部分，前段面向客人，可做备餐区，后段设有炉灶和抽油烟机，以应付较重要的烹调需要。

通过厨房区一端的趟门，可通到外面的庭园及车库。车库与庭园可互通，间隔墙刻意不搭建到底，在离地0.3 m处改用玻璃，制造出悬浮

效果。车库除供汽车停放外,靠墙的铁杆上还挂了两辆单车,颇具现代装饰的味道。影音室在车库隔壁,以黑、灰两色铺陈,墙壁和顶棚覆盖了隔音棉,地上也铺设了吸音的地毯,玻璃屏顶端的两道凹槽,分别隐藏投影幕和装设遮光帘。

The restaurant and entrance space are side by side with one wall as a partition. Excess garden partition wall is taken down. Green figure thus comes into the room. Designer creates an open kitchen, so that the vision is wider. TV background wall are designed very unobtrusive, by only modified square laminates, which successfully set off the chandeliers, tables and chairs. In fact, though the table and chandelier are made by solid wood or juvenile wood, it can be stand out from a piece of black, white and gray.

Kitchen and dining room form a Lu-shaped space. Tables, benches and cabinets are arranged in parallel in order to maintain walkways clear. Walls and cabinet are major white, creating a sense of space and brightness. Dining chairs, bar and bar stool are black as supporting roles. The kitchen is divided into two parts, the front faces the guests which can be a food prepare area. The rear section has a stove and a range hood to cope with more important cooking.

Walking through the sliding door beside the kitchen area, one can get to the outside garden and garage. Garage and garden are linked with each other. Partition wall is deliberately combined with 0.3 meters glass wall from the bottom, creating a suspension effect. In addition to the car park, the iron beside the wall are hung two cycling with a fairly modern decoration taste. Audio-visual room in beside the garage, using black and gray colors. Wall and ceiling are covered with insulation cotton, the floor is covered with sound-absorbing carpet. Two grooves at the top of the glass screen are respectively hiding projection screen and installing light shade.

# Marina Cove

匡湖居

设计单位：ANSON CHENG INTERIOR DESIGN LTD.
设 计 师：郑秋基
项目面积：180.6 ㎡
项目地点：香港
主要材料：大理石、铁器、木地板、砖

Design Company: ANSON CHENG INTERIOR DESIGN LTD.
Designer: Anson Cheng
Project Area: 180.6 m²
Project Location: Hong Kong
Major Materials: Marble, Iron, Wooden floors, Brick

业主繁忙的工作暂且告一段落，终于可以回到这个位于郊区的度假屋，放松紧绷多时的神经。将车子停好，打开大门步入大厅，光着脚踏在亚白色的地砖上，舒适自在的感觉立刻浮上心头。设计师规划的概念是要采用简洁、自然的风格，因此用色和材质运用上，以白色作基调，挑选具有天然质感的文化石、户外木等，增加在都市里度假的惬意度。另外，用通透的玻璃强调周边景观与光线的自然魅力，创造出淡雅、单纯、和谐的生活空间。

整个住所设计看似简单，但却不单调，单是色彩便做到日夜之别。在白天，鹅黄色的电视墙身，艳红的音响喇叭，以及淡黄绿色的单椅，能呈现属于年轻人的热情活力，展现温暖、明亮的气息。而当夜幕降临，文化墙身上方的LED灯槽可随需要亮起，灵活转变不同的彩度，适应各种情境，为现代、简约的空间交织富有层次的视觉美感。此外，线条的运用也以方圆形态互相平衡，例如家具有圆的凳子、方的沙发，又如灯具有圆的吊灯、方的灯槽，交融式的设计不只体现了线条美学，更有一种强烈的对比吸引力。

Owner's busy work has temporarily come to an end, and he can return to the suburban holiday homes to relax tense nerve. I park the car, open the door and step into the hall.My bare foot treads on the white floor tiles, comfortable feelingsoon comes up to the mind. Designer's plan concept is to use simple, natural style.So the use of colors and materials is based on white color, cultural stone and outdoor wood, etc., increasing the pleasant holiday feeling in

the city. In addition, transparent glass is stressing surrounding landscape and lighting natural charm, creating elegant, simple, harmonious living space.

Whole home design seems simple, but not monotonous, especially the color. During the day, light yellow TV walls, redaudio speakers, and light yellow-greensingle chair are able to show enthusiasm and vitality of young people, show warm and bright atmosphere. When night falls, cultural wall LED lights illuminates and changes flexibly under different saturation and various situations, creating visual beauty in the modern, simple spaces. In addition, the use of lines forms a balance of square and round shape, such as round stool, square sofa, round chandelier, square light tank. Blending style design not only reflects the aesthetic lines, but also forms a strong contrast appeal.

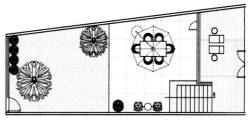
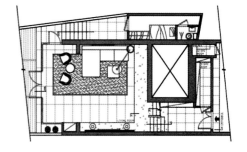
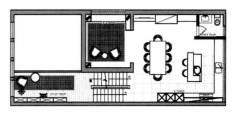
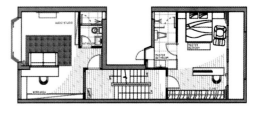

# Sausalito

岚岸

设计单位：Tint International Limited
设 计 师：冯建耀
项目面积：195 ㎡
主要材料：灰玻璃、大理石、黑钢框

Design Company: Tint International Limited
Designer: Eric Fung
Project Area: 195 m²
Major Materials: Gray glass, Marble, Black steel frame

本案有令人瞩目的玄关，在仰视香槟金的顶棚中，可看到摇曳动人的吊灯正悬挂在半空。玄关两旁铺上不同质感的材料，丰富了视觉画面。鞋柜的饰面运用扣皮铺贴，对面的客厅暗门用特色夹丝玻璃作修饰，与墙身连为一体。为塑造统一的感觉，选用香槟金钢框饰面点缀两旁，打造出时尚、不凡的气派。

跨过玄关走进客、餐厅，抬头可见雪白的假顶棚中勾勒着优雅花线框，灯光从花线旁渗透出温暖、舒适的柔和光线。电视墙上铺贴啡金色壁纸，黑钢框在灯光效果的映衬下，显得更加分明。

为满足客户的需求，设计师善用餐厅每处角落作为储物柜。不仅有镂空的饰品柜，亦有珍珠白扣皮饰面的矮柜收藏杂物，并在柜与柜之间留出空间放置心爱的钢琴，让女儿闲时在家中演奏一番。

为了将餐厅的视觉空间延伸，顶棚选用灰玻璃装饰，低调地反映出时尚魅力。素雅的壁纸与珍珠白色柜互相呼应，黑钢框的修饰，增添了设计层次，打造出个性的都市品位。大理石地台依循环境色系，继续以浅啡色延伸至客厅，烘托出古典的气息。

主卧迎合业主高贵的气质，利用不同深浅的啡色描绘出现代、经典的空间气质。啡金色几何形铺成的床头特色墙配以玫瑰色金钢枝修饰，突显出乱中有序。大理石墙面配搭香槟金钢枝，无形中加强独特的古典风味。顺着香槟金钢枝看过去，一组灰玻璃大衣柜展现于眼前，提升了空间的视觉效果。

The case have remarkable entrance.From the champagne ceiling, one can see moving swaying chandelier hangs in the air. Different sense of materials are put on both sides of the entrance, enriching the visual picture. Shoe cabinets'finishes is covered by leather. The opposed secret door uses special wire glass as modification, which links the wall. To create the feeling of unity, the designer uses champagne golden steel frame as ornaments, creating a stylish, extraordinary style.

Stepping across the entrance into the living room and dinning room, one can see white false ceiling outlining the elegant flower frame. Light penetrates warm, soft and comfortable lines from the frame. TV wall has golden brown wallpaper, so that the black steel frame under the light effect seems more clear.

To meet customer's demand, designer makes good use of the corner of the restaurant as storage cabinet. There is not only hollow jewelry cabinet, but also pearl white leather finishes holding things. Designer sets aside space for the beloved piano between the cabinets, daughter can play at home in some leisure time.

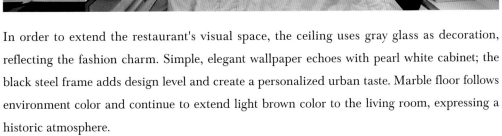

In order to extend the restaurant's visual space, the ceiling uses gray glass as decoration, reflecting the fashion charm. Simple, elegant wallpaper echoes with pearl white cabinet; the black steel frame adds design level and create a personalized urban taste. Marble floor follows environment color and continue to extend light brown color to the living room, expressing a historic atmosphere.

Master bedroom caters the owner's noble characters. Designer uses different shades of brown color to create a modern classical space. Golden brown geometric circular paved bedhead wallis decorated with rose gold steel sticks, highlights order in chaos. Marble wall matches champagne gold steel sticks, virtually adds unique classic flavor. Seeing along the champagne steel sticks, a group of gray glass large wardrobe enhances the space's visual effect.

# Oriental Bund B Unit

泷景 B 户型

设计单位：进思设计有限公司
项目面积：135 ㎡

Design Company: ELEVATION PARTNERS CO. LTD.
Project Area: 135 m²

本案为三房两厅，外加入户花园，空间强调连贯性和通透感的处理，活动的功能隔断也有助于增强空间的灵活性。

设计以"新亚洲精蕴"为主题，整体风格以简约为基础，以素色为主，暖色木墙身与浅灰色云石地台的搭配，营造出清新、自然的感觉，给家居带来一丝暖意。从客厅淡淡湖水色的壁纸，到家具的简约线条造型，再搭配东方韵味的饰品摆设，体现了属于业主的优质生活，贯彻了东方简约的美学精神。从墙上的剪纸艺术挂画中可感受到，设计师把佛山独有的岭南文化着墨于局部设计细节中，和谐地展现传统与现代设计项目的融合。

The case is three-bedroom-two-living room structure, with the home garden, emphasizing a sense of coherence and transparency. Movable functional partition helps to enhance the flexibility of space.

The theme of the design is "New Asia Essence". The overall style is based on simplify. The color is mainly plain. Warm wood walls and light gray marble floor are well matching with each other, creating a fresh, natural feel, bringing home a trace of warm. From the light-lake-color wallpaper to the furniture's simple line, as well as the oriental charm jewelry ornaments, they are reflecting the high-level quality of life belonging to the owner, and implementing the East minimalist aesthetic spirit. One can feel from the art of paintings on the wall, that the designer adds Foshan's unique Lingnan culture in the local design details, harmoniously showing the integration of traditional and modern design projects.

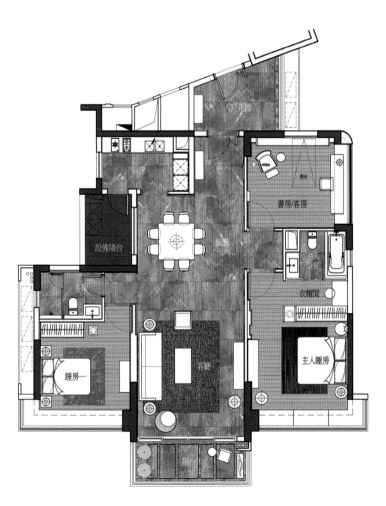

# Hill paramount

名家汇

设计单位：In Him's Interior Design
设 计 师：林俊谦
项目面积：204 ㎡
项目地点：香港
主要材料：大理石、木地板

Design Company: In Him's Interior Design
Designer: Samuel Lam
Project Area: 204 m$^2$
Project Location: Hong Kong
Major Materials: Marble, Wooden floors

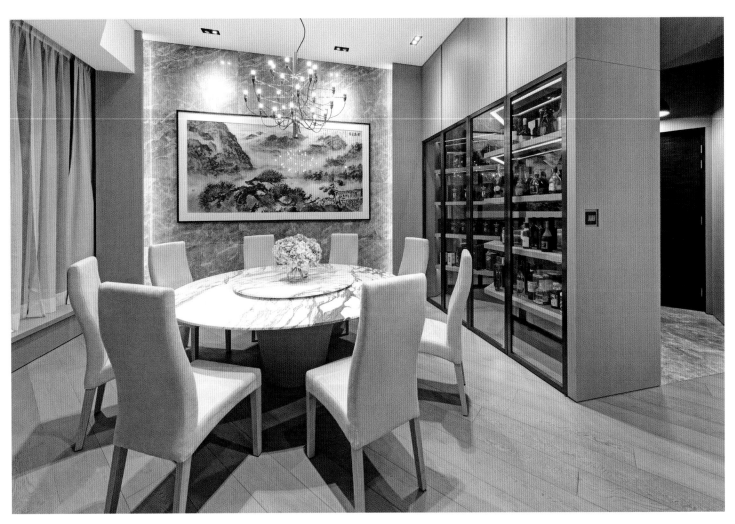

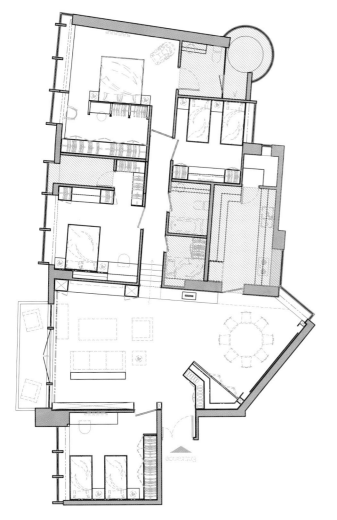

有别于一直沿用的设计概念，设计师并没有用大量的金色及亮面材料来烘托空间。本案采用了简单的设计手法，运用合理的材料搭配使空间散发出优雅气质。如房间的柜面以薰衣草灰色营造出丰富的质感。

另外，空间中加入了业主要求的大理石元素，最为显眼的就是餐厅的大理石墙及大理石餐桌，搭配一幅中国山水画，使整个空间简单中带有雅致的中式品位。

客厅则以惯用的简约格调为主，电视柜及沙发后面的架柜与华丽的餐厅形成对比。

私密空间则由气派变为低调、朴素，设计师用中性色调的材料打造出惬意空间，配以陈设品使人感受到优雅而舒适的气氛。

Unlike the long-time used concept, the designer has not very much golden and glossy materials to set off the space. The case uses simple design techniques and reasonable material makes the space exuding elegance atmosphere. The counter in the room uses lavender gray to create rich texture.

In addition, space is added into marble elements under the requirements of owner, the most prominent is the restaurant's marble wall and marble dining table.The Chinese landscape painting makes the entire space a simple, elegant Chinese style.

The living room is the usual minimalist style, TV cabinet and frame cabinet behind the sofa contrast with the ornate restaurant.

Private space changes from gorgeous to low-key, simple.Designer uses neutral materials to create a cozy space.The elegant furnishings make peoplefeelcomfortable and elegant.

# The Vineyard

葡萄园

设计单位：ANSON CHENG INTERIOR DESIGN LTD.
设 计 师：郑秋基
项目面积：321 ㎡
项目地点：香港
主要材料：木色地台、黑色壁纸、大理石地台、黑铁装饰

Design Company: ANSON CHENG INTERIOR DESIGN LTD.
Designer: Anson Cheng
Project Area: 321 m$^2$
Project Location: Hong Kong
Major Materials: Wood color platform, Black wallpaper, Marble floor, Black iron decoration

设计要突破传统、寻求创新，就要在细节、用色、层次及空间编排上花心思。玄关处的白色大门、黑色墙面，营造出丰富的层次。客厅以黑、白色作为设计主调，地台采用泥土色的砖石，偏厅的黑色墙面配以金黄色的画，形成了强烈的对比。虽然空间以黑、白色为主，但也重视不同材料之间的质感层次，空间中使用了白色大理石、钢琴漆及人造无缝石等，在不同的材料中营造出层次美。餐厅延续了黑、白色的风格，运用了不同形状的黑色灯罩，而内面则是闪耀的金色。定制的六张餐椅中，每个手工制作的餐椅的外形都独一无二。

上层空间主要以房间为主，采用木楼梯以增加温暖感。书房是开放式设计，在视觉上让空间变大，白色人造石的书桌与啡木色的楼梯融为一体的同时制造了扩大书桌的错觉。主卧以浪漫情调为主，以床尾吊板作为电视墙，解决了睡床与电视距离太远的问题，同时也分隔出了衣帽区及卧室的位置。

本案还设有一个私家花园，以不同的绿色植物作为盆景，再在旁边加设水帘瀑布，配以户外木桌及秋千藤椅，令业主感受到大自然的美。

If a design aim at breaking tradition and seeking innovation, it is necessary to pay attention on the details, colors, spatial arrangement. White gate, black wallat the entrance creates a rich level. Living room uses black and white color as the design keynote.Platform uses earthcolor brick, black wall and gold paint in the hall forms strong contrast. Although the space is in black and white, designer also pay attention to

the texture level between different materials. Space uses white marble, piano paint and artificial stone in different materials to create a tier beauty. The restaurant continues the black and white style, using different shapes of black lampshade, while the inner surface is shining gold. Among the customized six chairs, each one has an unique shape.

The upper space is mainly based on the room. Using wooden stairs to increase warmth sense. Study is an open design, so that the space is visually larger, white man-made desk and brown wood color staircase are integrated together, forming a sense of bigger desk. Master bedroom bases on romantic sentiment. The bed end hanging board is a TV wall, solving the problem of distance between the bed and TV, and also separating dressing area and bedroom.

The case also has a private garden, using different green plants as bonsai, and then adding curtain falls next to the bonsai. Outdoor wooden table and swing chair make the owner feel the beauty of nature.

# CHI on the Gardine's Lookout

气转乾坤

设计单位：P+P 室内建筑设计有限公司
设计师：廖奕权
项目面积：186 ㎡
项目地点：香港

Design Company: PplusP Designers Limited
Designer: Wesley Liu
Project Area: 186 m²
Project Location: Hong Kong

大门入口是很有格调的设计，地上铺设的是意大利雕花特色砖，烘托着啡色镜柜，墙边从顶棚垂吊了两盏小灯，向上可见一组两层的铜钱图案灯饰，材质及细节优美，线条明晰。大门的质感独特，进屋前已让人留下美好印象。沙发背后的白色鱼缸，成为了客厅生动的布景，鱼缸旁的白色装饰层架，摒弃了原本方方正正的沉闷排列，由不同大小的长方格组成，上下错落有致。部分层架采用玻璃，结合射灯效果，放置摆设后，装饰效果彰显出贵气。

客厅电视墙以木板装饰，上方设置了一块白层板，以珊瑚礁为摆设，在顶棚灯槽的渗光效果下，突显出立体感。而电视下方则放置了一块长木台，与白色焗漆储物抽屉相结合，长台下亦同时设有灯光。这样设计既让长台与储物抽屉看来轻盈，也与上方灯槽互相呼应，组成双线平衡的渗光效果，从而勾勒出电视墙框架。当灯亮着时，便有聚焦效果。

客厅旁边的餐区，与开放式厨房混合为一，让整体空间感觉较为宽敞。而厨房只有一部分橱柜及工作台与用餐区连贯，从橱柜一边再往前才是厨房的主要"腹地"。厨房内有一特色墙设计，白色的墙壁上不规则地分布着从墙壁凸出的小三角形层格，可摆放装饰，简洁而清新。而三角形层板的想法，是取自竹树上的叶，借以融合窗外的绿树环境。室内除了这个特色的墙壁外，儿童房的墙壁上方也用连绵的不规则的彩色储物方格作为装饰，整体形态如幻觉影像，既刺激了幼儿的视觉感官，又为空间增添了跳脱的线条。

The door entrance has a very stylish design, and the ground is covered with characteristic Italian carved brick, contrasting with the brown mirror cabinets.The ceiling is hanging two small lights, up to which are two layers of coins pattern lights, materials and details are beautiful, the lines are clear. The unique texture door leaves people a good impression before entering.

White fish tank behind the sofa has become a vivid scenery in the living room.White decorative shelf next to the aquarium abandons the original boring square arrangement, and it is composed of different sizes of squares from top to bottom. Part shelves are made of glass. The shelves and lighting effectshig hlight noble sense.

The TV wall in the living room are decorated with woodpanel, above which are set a white laminate. Coral reefs are displayedunder the ceiling light tank's seepage light effect, highlighting the three-dimensional effect. The bottom of the TV is placed with a long wooden table combining with white paint storage drawer. The lights are set under the table. The design allows long table and storage drawer seems light, and also echos with top light tank, forming a double-balanced light seepage effect, which outlines the frame of the TV wall. When the lamplit, it has

focus effect.

Dining area next to the living room is mixed with an open kitchen, so that the overall space is more spacious. Only part of the kitchen cabinet and table are connecting with dining area. Walking along the cabinets is the main "hinterland" of the kitchen. The kitchen has a feature wall, on the white wall are irregularly distributed small triangle layer lattice which can be used as decorative place, simple and fresh. The triangle laminates idea is taken from the bamboo tree leaves, to match the environment outside the window. In addition to the feature wall in the interior space, thewall of the children room are decorated with many irregular color storage boxes, the overall shape is like illusion image, which stimulates kid's visual sense, and also add jumping lines.

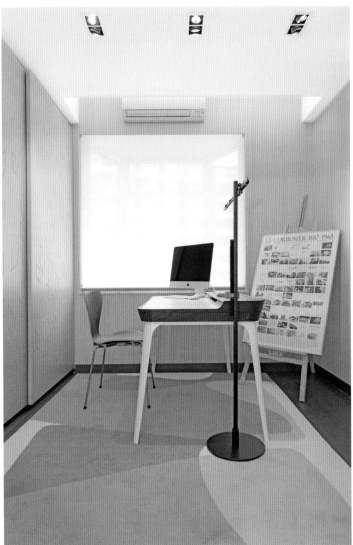
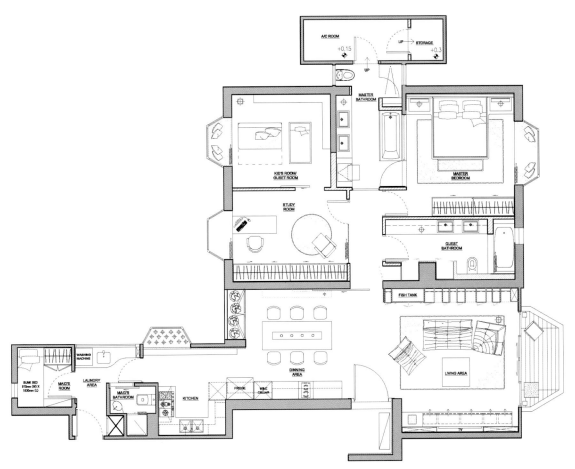

# The Rivapark

溱岸8号

设计单位：In Him's Interior Design
设 计 师：林俊谦
项目面积：186 ㎡
项目地点：香港
主要材料：大理石、木地板

Design Company: In Him's Interior Design
Designer: Samuel Lam
Project Area: 186 m²
Project Location: Hong Kong
Major Materials: Marble, Wooden floors

整个项目以简约、贵气为主调，一楼选用灰色大理石，配以铝制黑框的视觉点缀，为居室增添了现代元素，而不同材料之间的变化与层次，令整个空间在大方优雅之余，又不失别致的感觉。

客厅、餐厅运用了简洁的线条元素及黑色的家具，如电视柜钢架、黑玻璃餐桌及装饰墙等，不但为宽敞及以浅色为主的空间增添了视觉立体感，更展示了时尚气息。

随着木地板楼梯和黑色玻璃饰面墙身的延伸，设计师巧妙地把一楼的优雅、别致与二楼的简约风格连接起来，将以黑色作点缀的风格带入了二楼的公共空间。二楼主要为卧室，以白色及木材为主，打造出温暖的感觉。设计师在白色的空间里选用了使人倍感温暖的木质材料，以缓和白调的冰冷之感。主卧及两间卧室房间墙身上加了掩门，使各区域互相连接，不但能进行功能分区，又能保留空间联系，令空间运用更加灵活。

设计师巧妙地串联了雅致、简约、温馨的风格，大大提升了整个复式空间的连贯性。

The entire project is simple and noble, the first floor uses gray marble decorating with aluminum black frame, adding modern elements to the house. The change and layer between different materials make the entire space generous, elegant and unique.

Living room and dining room use simple lines and black furniture, such as TV cabinet steel, black glass dining table and decorative walls, etc., not only adds visual 3D sense for spacious and light color space and also adds fashion sense.

With the extension of wood floors and stairs and black glass walls, designer cleverly links the first floor's elegance and uniqueness with the minimalist style of the second floor. The black decoration style has been brought into the public space on the second floor. The second floor is mainly set the bedroom with white color and wood materials, creating a warm feeling.

Designer chooses warm wooden material in the white spaces to ease the cold feeling of the white tone. Master bedroom and two bedrooms' body has been added closing doors, so that all regions are interconnected. The space can not only remain functional partition, but also maintain the space connection and flexibility.

Designer cleverly links elegant, simple and cozy style, greatly enhances the coherence of the entire penthouse space.

# Tsing Yung Terrace

青榕台

设计单位：设计 2000
设 计 师：Charlie
项目面积：149 ㎡
项目地点：香港
主要材料：人造石、壁纸、灰镜

Design Company: Design 2000
Designer: Charlie
Project Area: 149 m²
Project Location: Hong Kong
Major Materials: Artificial stone, Wallpaper, Gray mirror

客厅、餐厅的墙面粗看是月白色,细心留意,却各有变化。四个卧室全部保留原有间隔,让三代的五位家庭成员都各自拥有相当自主的休憩空间。

客厅正对着青山湾,视野开阔,用栏门分隔了小露台,业主可随时踏足户外。由设计师定制的茶几,几面都是托盆,使用灵活,与影音柜一起强调了黑白对比。

主卧的套间浴室用清玻璃作隔墙,间隔顺着横梁规划,床尾空间做了大型衣橱。在主卧卫浴家具造型简单有型,设计师打造了挂墙的陈设架,令浴室有用于装饰点缀的好位置。洗手台以方形台盆为主角,L形的墙面全铺设清镜,既方便梳洗,又可把柱子加以修饰。

男孩的卧室铺设了升高一级的木地台,用深啡色的橡木,衬以白色衣柜。衣柜拉门用横条子灰镜装饰,避免显得笨重。在窗前摆放了书桌,墙上装设的层架是业主的要求,但采用红、黑衬色则是设计师特别为年轻人所打造的,效果出奇地受欢迎。

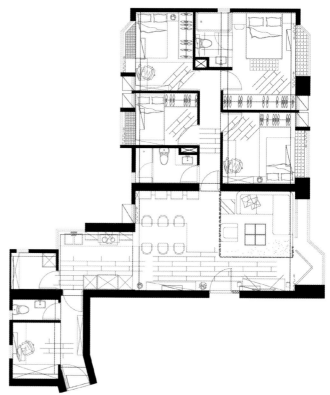

The living room, restaurant's wall roughly looks like white. If you pay careful attention, they have slight difference. Four bedrooms retain the original interval, letting the five family members of three generations have quite independent open free space.

Living room is facing the Green Mountain Bay. The slight is wide. The door separates the small terrace, the owner can set foot outdoors at any time. Coffee table customized by the designer has four sides pallet around. The use is flexible, forming black and white contrast with the audio-visual cabinet.

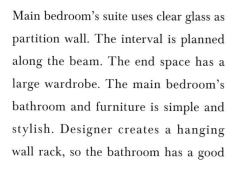

Main bedroom's suite uses clear glass as partition wall. The interval is planned along the beam. The end space has a large wardrobe. The main bedroom's bathroom and furniture is simple and stylish. Designer creates a hanging wall rack, so the bathroom has a good location for embellishment. Washbasin has an outstanding square basin, L-shape wall has clear mirror which is convenient for grooming and modifies the column.

Son's bedroom is covered with an elevated wood table which is made of dark brown oak and white wardrobe. The wardrobe door is decorated with gray stripe mirror so that the door will not seem too heavy. The owner wants to put a desk in front of the window, and install rack on the wall. Red and black color are created by the designer for young people, the effect is surprisingly popular.

# Sorrento

擎天半岛

设计单位：COMODO Interior & Furniture Design CO.,Ltd.
设 计 师：王智衡
项目面积：158 ㎡
项目地点：香港
主要材料：灰色木地板、灰色壁纸、灰镜

Design Company: COMODO Interior & Furniture Design CO.,Ltd.
Designer: Alain Wong
Project Area: 158 m$^2$
Project Location: Hong Kong
Major Materials: Gray wood, Gray wallpaper, Gray mirror

业主喜欢设计师一贯的风格：设计以时尚、简约及温馨为方向，以深浅变化的灰色与白色混搭，加上带有灰色调的木地板及不同材料的混合，在丰富触感之余，同时为空间增添温馨的暖意。

将客、餐厅与旁边房间的墙拆掉，改用玻璃间隔，打造一个书房空间。沙发旁及书房内选用了相同的灰色壁纸，令两个空间更为连贯，使视觉不受局限。沙发背后做了半腰高的地柜，恰当地增加台面及储物空间。落地玻璃屏及隐藏门令线条更简洁。

玄关附近有一个鞋柜，柜门用了灰镜装饰，在视觉上扩宽了入门空间。餐厅的布置尽量简单，白色较多，平衡了整个空间的灰色调。木饰面一直延伸至鞋柜侧，那里暗藏了一个对流窗，避免墙面太多断口，有助于视觉统一。装饰镜屏与黑钢框的组合，跟书房玻璃屏互相呼应。

除了客厅与书房的灰色壁纸,电视背景墙也用了米白色壁纸,看似油漆,实有细致的触感。定制的家具表面除了常用的木皮外,部分改用亮面焗漆,如电视柜、餐厅半腰柜、书房柜组等,书房桌面更采用玻璃材质,增加其光亮感和时尚感。

设计师恰当地运用主卧门后的空间,除了腾出空间建造衣帽间,相对扩大了主浴室,这与一般在床尾摆放一列高柜,清楚划分区域的做法相比,避免了睡眠区的压迫感。床头板简单采用了黑色绒布软垫,与灰、白两色形成对比。床侧的衣帽间间隔墙面用木饰面铺贴,划出了宽度不一的条纹。主浴室与衣帽间相连,以磨砂玻璃趟门分隔,除了有足够的替换空间,也符合业主日常使用的习惯。

厨房墙面及地台都采用灰色瓷砖,还有一面墙面采用了不锈钢面板。橱柜用木纹胶合板及亮面焗漆,带来时尚又不失自然的气息。

其他两个房间以衣柜分隔,一个是儿童房间,另一个是后备房间。考虑长远规划,儿童房间以白色为主色调,适当地配衬粉红色,比较耐看。

Owner likes the designer's usually style: design direction is fashion, simplicity and warmth. The gray and white color are mixed with each other. The mixture of gray wood floor and other materials add a warm feeling to the space with a rich touch sense.

Designer removes the wall of living room and dining room and uses glass partitions to create a study. Sofa and study have same gray wallpaper, so that the two spaces are more coherent, and the vision has no limitations. Behind the sofa there are a half-waist high cabinets, appropriately increases the table top and storage space. Ground glass screen and hidden door make the line more concise.

There is a shoe cabinet near the entrance, which is decorated with gray mirror, visually widens the entry space. The restaurant's layout is as simple as possible. The white color balances the gray color in the whole space. Wood finishes are extended to the shoe cabinet, where hides a convection window, avoiding too much fracture walls and help for visual unity. Combination of decorative mirror screen and black steel frame is echoing with the glass screen in the study.

In addition to the gray wallpaper of living room and study, TV background wall uses beige wallpaper, which looks like paint, and actually it has delicate sense of touch. Except for usual wooden cover, designer uses light paint on TV cabinet, restaurant half-waist cabinet, study cabinet, etc.. Study desktop is made of glass, increasing its brightness and fashion sense.

Designer appropriately uses of the space behind the door of the master bedroom. In addition to make room for the construction of the cloakroom, designer expands the master bathroom. Comparing with the traditional way of putting a series cabinets at the end of the bed while clearly divide the space, such way avoid sleep pressure. Headboard is made of simple black velvet cushion, which is greatly different from gray and white color. Cloakroom's interval side wall uses wood finishes, drawing different wide stripes. Master bathroom and cloakroom are connected with separation abrazine glass door, except for enough space to replace, it also in line with owner's everyday use.

Kitchen wall and floor tiles are gray, another wall is made of stainless steel panel. Cabinets are using wood veneer and light baked paint, bringing stylish and natural atmosphere.

The other two rooms are separated with wardrobe, one is for kid room, the other is for reserved room. Considering long-term planning, kid room is white color with appropriate some pink, the room is good looking.

# Yihe Mountain & Lake House

颐和湖光山舍

设计单位：上上国际（香港）设计有限公司
设 计 师：曾宪明
项目地点：海南三亚

Design Company: Shang Shang International (HK) Design Co., Ltd.
Designer: Zeng Xianming
Project Area: Sanya, Hainan

本案采用具有现代感的亚热带海岛城市休闲度假风格，设计元素取自于阳光沙滩，运用现代简约风格表达创新理念，设计师在不同空间场景从不同角度诠释空间的优雅气质。软装配饰以海的元素为主，选取了自然材质，彰显出文化艺术特色。

观景阳台让美景尽收眼底。大客厅使阳光、鲜氧和煦流连。雅致飘窗可让人在此品茗静思。开阔的格局让客厅、餐厅、厨房连接成一体。空间中感受到木质的温润、阳光的柔和，简雅画作带来一丝丝艺术气息。设计师在本案中遵循了"Less is more"的设计理念，表现出阳光沙滩的简约美学。

"Less is more"倡导的是一种极简主义、一种减法原则。在美学范畴里的极简，其内在远非表现手法上的节俭，而在于一种克制的精神，简单的东西往往带给人们的是更多的享受。

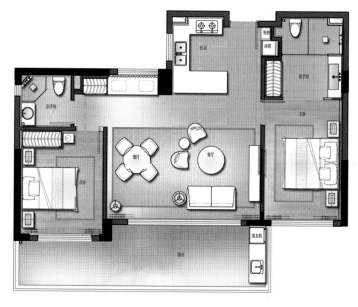

The case is modern urban sub-tropical island leisure style. Design elements are taken from sunshine and beach, using modern, simple style to express innovative ideas. Designer uses different spatial scene and angles to interpret elegance mood. Soft decoration bases on sea elements, the natural materials highlights the artistic culture.

View balcony allows a panoramic view. Large living room uses sunshine, fresh oxygen and warm wind. Elegant window can make people here drinking tea and thinking deep. Open pattern links the living room, dining room, kitchen together. The space is full of warm wood, gentle sunshine, a hint of art brought by simple but elegant paintings. Designer follows "Less is more" design philosophy, showing the minimalist aesthetics of sunshine and beach.

"Less is more" advocates a minimalist principle, a subtraction principle. The minimalist in the aesthetic category is not internal simple, but a restraint spirit, simple things often bring people more enjoyment.

# House on the Peak

山顶大宅

设计单位：Danny Cheng Interiors LTD.
设 计 师：郑炳坤
项目面积：836 ㎡
项目地点：香港
主要材料：镜钢顶棚、白玻璃、大理石

Design Company: Danny Cheng Interiors LTD.
Designer: Danny Cheng
Project Area: 836 $m^2$
Project Location: Hong Kong
Major Materials: Mirror steel roof, White glass, Marble

楼高7层的山顶大宅，配以时尚、独特的内外造型，给人一种亲切的时尚感。大门口以镜钢顶棚、白玻璃等反光材料装饰，除了能增强空间感外，还能给人一个时尚及有气派的第一印象。

这间住宅拥有先天的优点：下层偏厅的楼顶高，拥有充足的空间感。设计师为避免偏厅令人感到大而空荡，所以特意以不同的材料及布局，增加空间的层次，突显空间上的变化。例如，把上层客、餐厅与偏厅间的间隔墙打掉，用感觉更轻盈更纤薄的"龙骨楼梯"，让上、下两厅能相互连接，使空间在视觉上更为宽阔，以及增加了不同空间之间的沟通。余下的墙面部分改为黑色，就像画上一个"7"字，在刻意分割两厅之余也带一点个性。上层客厅、餐厅以白色为主色调，令业主感到轻松、舒适，能够舒缓日常工作中的紧张情绪。客厅中深色地毯用以缓和过度纯白的冰冷感，以及连贯下层偏厅的用色。客厅的一旁是餐桌位置，设计师没有采用豪华夸张的大号餐桌，而是用了简洁、线条感强、与地下用同一材料的雅士白餐桌。客、餐厅看上去没有多余的柱子，这全靠设计师把柱子隐藏在设计中。

室内的设计无论内外均经过了精心的布局，这样才营造出一个时尚、含蓄，散发品位的山顶大宅。

Seven-storey summit house with fashionable, unique styling inside and outside, giving people a sense of fashion. The entrance has mirrors steel roof, white glass and other reflective materials as decoration, enhancing the sense of space, and giving people a stylish and lord first impression.

This house has inherent advantages: the lower slant room has high roof and plenty of space. To avoid making people feel over big and empty of the slant room, designer deliberately uses different materials and layout to increase the space tier, highlighting the change of space. For example, removing the partition walls between the upper floor living room, dinning room and slant room, using lighter and slimmer "keel stairs" to link the upper and lower halls, making the room visually wider, and adding communication between different spaces. The remaining part of the wall can be changed into black, just like painting a number "7", deliberately sets the two halls apart. Upper living room and dinning room are in white tone, making owner feel relaxed and

comfortable, relieving tension in his daily work. Besides, the living room's dark color carpet is used to moderate the pure cold sense and to connect the lower slant room's color. Beside the living room there is a table. Designer does not adopt luxury and exaggerate big table but uses simple, strong-line-sense Aston white table. Living room anddinning room have no extra columns, because the designer hides the pillars in the design.

Both interior design and exterior design are through careful layout, so that creating a stylish, subtle taste summit house.

# Bellagio

深井碧堤半岛

设计单位：COMODO Interior & Furniture Design CO.,Ltd.
设 计 师：王智衡
项目面积：93 ㎡
项目地点：香港
主要材料：木纹地砖、壁纸、定制家具

Design Company: COMODO Interior & Furniture Design CO.,Ltd.
Designer: Alain Wong
Project Area: 93 m²
Project Location: Hong Kong
Major Materials: Wood floor tiles, Wallpaper, Custom furniture

设计师 Alain 对家居的要求很简单：不需要有复杂的装饰，希望有一个能令他心情完全放松，让他流连忘返的舒适家居。Alain 平日工作繁重，一年中会与太太外出旅游数次，目的地都是看得到大海的地方。Alain 说："不一定只有平静的海面才能给予人安稳的感觉，大海在惊涛骇浪后归于宁静反而令人心情平静，感觉有如洗涤心灵，有时候甚至会刺激思考，使人得到新的灵感。"因此，购入这个住宅时一望无际的海景是考虑的因素之一。Alain 将沙发摆放面向客厅的窗户，同时在窗台上加上软垫，悠闲时可以自在地看着大海的变化。

Alain 的设计着重线条和比例，还有不着痕迹地修饰每个细节。客厅中大片深啡色及白色的木皮墙面交接，尽量将墙壁连贯起来，减少墙面分割，在视觉上有助于扩展空间。沙发旁的墙面贴上深啡色木皮，由窗边一直延伸至走廊末端，修饰了主人房的房门。而白色木皮墙身则隐藏了客房及客浴室门，延伸至位于客厅另一端的开放式厨房。

设计师偏好柔和、浅色及木材，住宅内采用带点灰调的啡色木纹地砖，餐厅摆放木材餐桌及长椅，主卧床头饰面板及主浴室活动趟门，搭配浅灰色和米色的壁纸，还有雪白的定制家具，整体设计现代、简约，又不失自然、舒适。

由于原有的餐厅设计并没有窗，减少了自然光照，令餐厅比较暗。设计师特意在近大门的墙面镶上了一幅清镜。通过反射作用，把客厅的自然光引入餐厅，令空间更加光明、洁净。

拆去书房与主卧的间隔墙，将两室贯通。利用窗台改造成书台，窗外漂亮的海景提供了一个舒适轻松的工作环境。主浴室选用趟门隐藏在墙面之中，提升主浴室的美观感。卧室用了较深的色调，搭配有丰富质感的材料，如布窗帘及绒面软布床头饰面板，营造出一个舒适的空间，提升了空间中的亲密气氛。

Designer Alain has a simple home request: a place where has no complicated decoration, but can make him completely relaxed and comfortable. Alain has a heavy workload every day, he will travel with his wife several times in a year and most of the destination are sea. Alain said: "Not only the calm sea can give people calm feeling, tranquility after ocean's tempestuous waves can also give people the same feeling which cleans the soul and sometimes even stimulate thinking, making people get new inspiration." So the vast expanse sea view is one of the factors when purchase the house. Alain puts sofa towards the living room window, while adds cushions on the window, one can freely see the change of the sea at leisure time.

Alain's design focuses on lines, proportions and detail modifications. A large area of coffee color and white color wood cover walls are connected. The consistent wall can reduce the wall

division and visually good for expanding space. The pasted coffee wood cover has been extended from the window to the end of the corridor, modifying master bedroom's door. The white walls hide the guest room's door and guest bathroom's door, extending to the open kitchen on the other side of the living room .

Designer prefers soft, light color and wooden materials; house uses coffee wooden floor tiles with a little gray tone. The dining room has wood tables and benches. The main bedroom's bed plate and main bathroom's door use light gray and beige wallpaper, and the white custom furniture seems modern, simple, but natural, comfortable.

Since the original restaurant design has no windows, therefore reduces the natural light, making the restaurant dark. Designer deliberately installs a pair of mirror close to the door wall. By reflex action, the mirror leads natural light from the living room to the restaurant, making the space more light and clean.

Designer removes the partition wall between the study and master bedroom, making the two chambers linked. Then he transforms windowsill into book desks, beautiful ocean views outside the window provides a comfortable and relaxing working environment. The main bathroom uses sliding doors that could be hidden into the wall to enhance the aesthetic sense of the master bathroom. The bedroom uses dark color matching with rich texture material, such as cloth curtains and soft cloth bedside decorative panels, creating a comfortable space, enhancing the space intimation atmosphere.

# Type B Apatment
# La Couronne

天御豪庭 B 户型

设计单位：于强室内设计师事务所
设 计 师：于强
项目面积：180 ㎡
项目地点：广东深圳

Design Company: Yuqiang & Partners
Designer: Yuqiang
Project Area: 180 m²
Project Location: Shenzhen, Guangdong

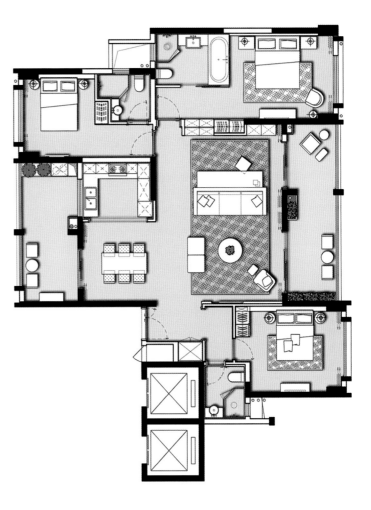

本案所定位的时尚现代风格有着简约、明快的基调，简练、精致的线条是空间立面的主体，具有品质感的经典款家具烘托出低调、华丽的氛围。材料采用亮光漆木饰面、丝绒、高档的壁布与大理石，细节有精致黑砂钢踢脚线与色彩丰富的地毯，运用 Tom Dixon、La Petite Voliere、Innermost 的名贵灯具和 Ibride 的 "公子的秘密" 书柜，装饰采用金属质感与晶莹剔透的饰品，体现出浓郁的时尚现代气息与文化艺术格调，雅致却不失亮点，将业主领入一个高端、现代的空间氛围中。

The case's fashion, modern style has simple and bright tone. Concise, elegant lines are the facade's main part. The classic furniture expresses a low-key, gorgeous atmosphere. Materials are light lacquered finishes, velvet, high-level wall covering and marble, with fine black sand steel skirting and rich carpets. It uses Tom Dixon, La Petite Voliere, Innermost's luxury lamps and Ibride's "prince's secret" bookcase. Decorations are using metal texture and sparkling ornaments, reflecting the rich fashionable modern style, which seems elegant and outstanding, bringing the owner into a high-end, modern space.

# Chatham Gate

升御门

设计单位：进思设计有限公司
项目面积：250 ㎡
项目地点：香港

Design Company: ELEVATION PARTNERS CO. LTD.
Project Area: 250 m²
Project Location: Hong Kong

本案为现代简约风格，运用几何形式的线条、编织细致精巧的布料、索色处理的木材、纹理独特的大理石，以及时尚、独特的家具来彰显这份非凡气派。

进入室内，首先被顶棚上的水晶吊饰所吸引，配合眼前偌大的玻璃窗外，那里同样充满了几何造型新建筑物的繁华景色，此时水晶吊饰仿如一片水晶浮云。晚上，水晶灯配合建筑物的璀璨灯饰，迷人夜色使人目不暇接。在客厅窗对面，摆放有电视，由于这面是主体墙，为了安排视听影音器材，并使立面整齐、简洁而不致零碎、混乱，设计师以暗藏手法处理，材料选用灰镜及灰镜钢，并且用经过索色处理的木材来衬托，使整体空间井然有序。

走进各房间之前，可观察到通往各房间的一道走廊比较狭长，设计以镜面装饰墙面，把狭窄的空间视线打开，再巧妙地把客厅云石地台伸延至走廊中段。在镜子的反映之下，把原本狭长的走廊缩短，同时也让客厅的空间视线得以伸展和延续。本案的四个房间，分别是客卧、儿童房、书房及主卧，房间以暖色调为主。其中，把书房安排在儿童房与主卧之间，并加设趟门，使书房可同时分别给两边使用，也方便父母与孩子一起做功课，是一个多功能的亲子书房。

主卧内包括衣帽间及浴室，在材料的选用上与客厅相互呼应，而色调以暖色为主。原建筑衣帽间较小，又是三角形，设计师采用玻璃将衣帽间重新设计，使空间实用性增强。主卧采用地毯铺设，营造舒适感。地毯的图案设计时尚、新颖，以立体几何造型为主，跟客厅中的几何图案互相辉映。

整体设计以娴熟的设计手法处理，线条简约、新颖独特，对空间的改善也给人留下深刻印象。

The case is modern minimalist style. Designer uses geometric lines, delicate fabrics, plain color wood, unique textured marble, and stylish, unique furniture to highlight the extraordinary style.

Entering into the room, one will be first attracted by the crystal drop on the ceiling. Out of the huge glass windows in front of the eyes, there is full of geometric buildings and bustling landscape. At this time, the crystal drop is like a piece of crystal cloud. On the evening, crystal lamps and bright lighting of buildings and charm night views make people dizzying. Opposite the living room window, there is a TV set. Since this is the main wall, in order to arrange audio-visual equipment, and make the facade neat, simple and without fragment and confusion, the designer uses hidden tactics to deal with the design. He selects gray mirror, gray mirror steel and plain color wood to make the overall space organized.

Before entering into each room, there is a relatively narrow corridor leading to each house. Designer uses mirror to decorate the wall, open the narrow space sight. And then he skillfully extends marble floor of the living room to the middle of the corridor. Under the reflection of mirror, designer shortens the original narrow corridor, meanwhile expands and continues the living room's vision. The case's four rooms are relatively guest bedroom, children room, study and master bedroom. The room bases on warm tones. The study is arranged between the children room and master bedroom, adding a sliding door, so that the study can be used on both sides. Therefore it is convenient for parents and children to do homework together, it is a versatile children-parent study.

The master bedroom includes a cloakroom and a bathroom. On the selection of materials, it is echoing with the living room. The color tone is warm. The original building's cloakroom is small and a triangle shape. Designer uses glass to redesign the cloakroom, enhancing the practice of space. The master bedroom is covered by carpet, creating a comfortable feeling. Carpet design pattern is fashion, new, and mainly three-dimensional geometric model, echoing with the living room's geometric pattern.

The overall design uses skillful design approach. The unique, new, simple lines and changes give people a deep impression.

# Gramercy
# A Unit

瑧环 A 户型

设计单位：进思设计有限公司
项目面积：118 ㎡
项目地点：香港

Design Company: ELEVATION PARTNERS CO. LTD.
Project Area: 118 m²
Project Location: Hong Kong

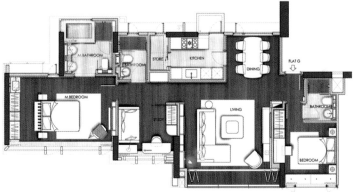

本案地处港岛中半山，毗邻苏豪区，拥有一个大家庭的业主，追求的是稳重、低调的豪华生活。业主具有高格调的品位，是城市中含蓄、奢华生活的拥有者。设计灵感源自品位颇高的精品酒店。整体空间采用了深咖色调，以胡桃木为主材，配以皮革，部分家具做钢琴面烤漆处理。利落的线条和精准的灯光配置，使空间具有高雅的格调。家具和摆设，均经过设计师的悉心构思，如客厅的大型现代水墨画，提升了空间的艺术品位。

The captioned case is located at Mid-levels Central in Hong Kong, next to Soho. The owner who has a big family, aims at pursuing a stable, low-key, luxury life. The owner has high-quality taste, he is the owner of the city's subtle, luxury life. The design inspiration comes from high-quality Boutique Hotel. The whole space bases on dark brown tone, and uses walnut as the main material. Besides, designer uses leather to match the walnut, and part of the furniture are made with pianos surface. Clean lines and precise lighting configuration gives the space an elegant style. Furniture and furnishings are all carefully designed by the designers, large-scale modern ink painting enhances the artistic quality of space.

# Hermitage

帝峰皇殿

设计单位：深圳市林文学装饰工程有限公司
设 计 师：林文学
项目面积：114 ㎡
项目地点：香港
主要材料：仿大理石砖、水曲柳木、特色壁纸

Design Company: (ShenZhen)Man Lam Interiors Design Co.,Ltd.Engineering Co., Ltd.
Designer: Man Lam
Project Area: 114 m$^2$
Project Location: Hong Kong
Major Materials: Imitation marble tiles, Ash wood, Features wallpaper

本案坐落于香港大角咀，一个旧区重建的边缘地带。根据客户要求，需要营造一个具有温馨格调的家。要改造这套房子，设计师面对着一定的挑战。一走进室内，入户门直冲着阳台，客人可以一眼看透客、餐厅，虽有广阔的视野，却无法让视线停留在精心打造的空间内，所以设计师用了一扇屏风作隔断，既可规划出玄关位置，也可轻微透视室内的装饰。客厅在装饰上采用了天鹅绒面硬包、定制的地毯等，增加了室内温馨的感觉。

娱乐及工作区以玻璃作间隔，融入客厅，使客厅感觉变大，亦减少了过道太长的压抑感。男孩房跟活动室分别在过道的两边，以玻璃隔屏让视觉贯穿，一致的布局及色调，将分隔两旁的不同空间融合，让孩子的领域得以扩展。主卧房亦将原有隔墙拆除，使两个小房间变成一个套房，既有回家的感觉，亦有酒店的格调。

值得一提的是，香港高楼的房子，窗台较多。设计师巧妙地将所有窗台，都变成室内空间来使用，或是休闲的聊天软座，或是床铺，或是工作台，使房价高昂的蜗居得到充分使用。

The case is located at Tai Kok Tsui in Hong Kong——an edge place of a renewed urban. According to customer's requirements, they need a warm style home. To transform the house, designer faces certain challenges. When entering into the interior, the entrance door is straightly open to the balcony. Guests can see

through the living room and the restaurant at one glance. The space has board perspective but the sight can not stay in it, so the designer uses a screen as partition. The partition can plan out the entrance location and also slightly disclose the interior decoration. The living room decoration uses velvet surface hard pack, tailor-made woven carpets which increases the indoor warm feeling.

Entertainment and working area use glass as partition. Therefore the living room space seems larger, and it reduces the sense of oppression of the long aisle. Boy room and activity room are at both sides of the aisle. The glass partition allows the vision through the space. Consistent layout and color integrate the different spaces on both sides of the partition, so that the kids field can be expanded. Designer removes the main bedroom's original existing wall to make two small rooms into a suite. The house has a homey feeling and also the hotel's style.

It is worth to be mentioned that, Hong Kong's high apartments have many windows. Designer skillfully changes all the window into interior space, or a casual chat soft seat, or a bed, or a bench, making full use of the high price houses.

# Beacon Heights

毕架山花园

设计单位：Millimeter Interior Design Ltd.
设 计 师：廖子扬
项目面积：120 ㎡
项目地点：香港

Design Company: Millimeter Interior Design Ltd.
Designer: Michael Liu
Project Area: 120 m$^2$
Project Location: Hong Kong

这间120㎡的公寓位于香港九龙北部的高档住宅区笔架山。门口设计了一个多用途的木质长椅。走廊巨大的镜子有效地将空间狭窄之感一扫而空。一进入起居室，立即会被一堵散发着芬芳而有益健康的气息的"绿墙"所吸引。这个绿色主题的设计可以使业主在久居城市的同时，还能享受到自然的眷顾。墙上种有红边竹蕉、袖珍椰子、地星、红掌等植物。

而全自动灌溉系统的安装可以使绿墙的日常护理更加便捷。此系统可以自动调节浇水的时间，从而使每种植物都可以吸收生长所需的适量水分。

木纹较深的阿拉莫橡木被选作起居室的地板用木，完美地与鸟笼地灯搭配在一起，使人看到后一种愉悦感油然而生。厨房的餐桌也选用了木质材料，桌上的东南亚风格台灯与主题风格完美贴合。主卧中的开放式盥洗室由玻璃围成，这个简约而雅致的设计不仅增大了原有空间，还使业主在浴缸中能够边洗澡边看电视，实在是一种享受！

This 120 m² apartment is located at Beacon Hill in north Kowloon, Hong Kong, it is a top-end residential area. At the entrance, there is a multipurpose wooden bench. Corridor huge mirror effectively sweeps away the narrow sense of space. When stepping into the living room, one will be attracted by a "green wall" with fragrance and healthy flavor. The green theme is designed to enable the owner enjoy nature's blessing while live in the city at the same time. On the wall are red edge bamboo banana, pocket-sized coconut, ground star, anthurium and other plants.

The installation of automatic irrigation systems can make green wall's day care more convenient. This system can automatically adjusts watering time, so that each plant can absorb the right amount of water for growth.

Alamo darker oak wood is selected as the living room floor wood which perfectly mix with the cage lights, a sense of joy wells up after people see it. Kitchen table also selects wood materials. The southeast Asian style table lamp perfectly fits with the theme style. The open master bedroom is enclosed with glass, this simple and elegant design not only increases the original space, but also makes the owner to has a tub shower while watching TV. This is really a pleasure!

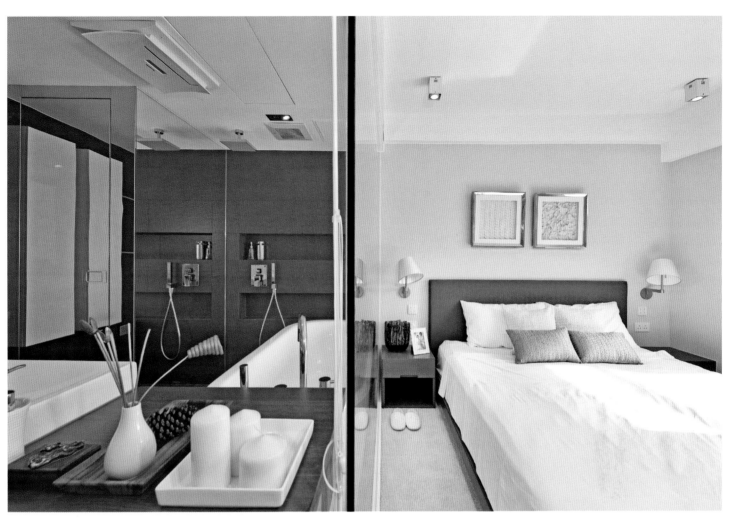

# The Beverly Hills

比华利山别墅

设计单位：深圳市林文学装饰工程有限公司
设 计 师：林文学
项目面积：300 ㎡
项目地点：香港
主要材料：意大利灰大理石、实木复合地板、仿大理石砖

Design Company: (ShenZhen)Man Lam Interiors Design Co.,Ltd.
Designer: Man Lam
Project Area: 300 m²
Project Location: Hong Kong
Major Materials: Italian gray marble, Parquet, Imitation marble tiles

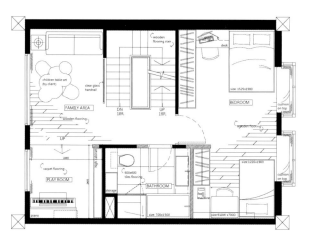
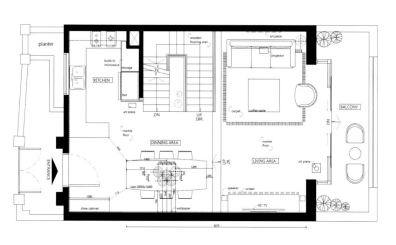
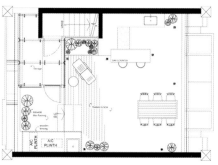
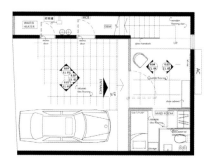

本案从地下车库到天台，共5层，地处香港，可谓是豪宅。从站在车库的入户门前开始，已经感受到高雅的格调，全黑镜面配皮革大门，彰显出气派。进入室内，可看见一面全身镜、一个鞋柜、几幅挂画、一张黑色的欧式椅子，虽然简单但很有味道。

沿楼梯走上客厅，映入眼帘的是意大利灰大理石地面，搭配简约的家具，既有高雅的气息，又不会眼花缭乱。餐厅里金光闪闪的吊灯，为空间增加了几分贵气。由于业主平常很少招待客人来访，所以，设计师也没刻意去堆放太多的沙发椅子，反而使室内有充足的空间感。再踏上楼梯，天然实木的梯步，顿时让人忘却大理石的冰冷，有一种温暖的感觉。

到了二楼，这里是孩子的天地，房间布置活泼可爱。过道是一面黑镜，不一样的是，设计师刻意把这面黑镜做了磨砂，可以当黑板使用，让小主人在这画图写字，尽情发挥创意。余下的空间，是亲子乐园，父母陪伴小孩一同玩乐，可以弹弹琴、唱唱歌，空间氛围甚是惬意。三楼是主卧套间，摆放有马毛饰面的屏风，将电视和衣帽间稍稍隔开，区域分布清晰，亦不失整体的统一感。

按业主要求，独立做了一个视听室，用作听歌及阅读之用，突显出业主的品位。在露台上，设计师刻意地多做几级梯步，并把天台大半部分地面升高，用意是将人站立时的视线，提升到高于原建筑围墙的高度，使人面对着海景树林，视野能够一望无际，心情变得豁然开朗。

The case has five floors from underground garage to the rooftop. It is located at Hong Kong and a luxury house. Standing at the entrance of the garage, one can feel the high level style. The whole black mirror and the leather door are generous and luxury. Stepping into the door, one can see a full-length mirror, a shoe cabinet, a pieces of painting and a black European chair. Their design are simple and charming.

Stepping to the living room along the stairs, the first thing comes into eyes is an Italian grey marble floor. The simple furniture has noble sense and will not make people dazzling. Glittering ceiling lamp adds sort of noble sense to the space. Since the owner usually seldom treats guests, designer does not put too many sofa and chairs in the space, therefore it has many empty rooms. When stepping on the stairs, natural wooden stairs make people forget the cold of the marble and feel a warm feeling.

The second floor is the kid's world. The room layout is lovely. Aisle is a black mirror, while it is different from normal black mirror since the designer deliberately does abrazine on it, therefore it can be used as a blackboard. The kid can write and paint at there with his creativity. The remaining space is a play room in which parents accompany children together to play the piano and sing songs, and the space atmosphere is very pleasant. The third floor is the master bedroom suite, where has horsehair screens separating the TV and cloakroom. The zones are distributed clearly with overall sense of unity.

According to the requirements of the owner, designer makes a separate auditorium, used for listening to music and reading purposes, highlighting the owner's taste. On the terrace, the designer deliberately makes more steps and elevates parts of the terrace floor. The intention is to enhance the people standing sight to the height of the original building walls. People are facing the sea view woods, the vision can be endless, the mood becomes bright open.

# The Joyful Tree House

半山壹号

设计单位：P+P 室内建筑设计有限公司
设 计 师：廖奕权
项目面积：148.6 ㎡
项目地点：香港
主要材料：木、激光切割木纹木质壁画、生态壁纸、磁性壁纸、镜

Design Company: PplusP Designers Limited
Designer: Wesley Liu
Project Area: 148.6 m²
Project Location: Hong Kong
Major Materials: Wood, Laser cutting wood wooden mural, Ecology wallpaper, Magnetic wallpaper, Mirror

进入电梯等候厅，映入眼帘的是不锈钢隔板，侧面有一个带镜鞋柜。当你体验着木地板的温暖感觉时，看到用激光切割木纹的木质壁画时，此刻，你感受到的不仅仅只是一处现代化的住所，而是一处独具特色的奢华舒适的家园。餐厅流露出古典与现代的混搭风格。顶棚上装饰着Tom Dixon设计的灯具，光影映射着白色的生态壁纸，就像穿透丛林树叶的阳光一样，创造了一种坐落于茂密丛林中的现代家园的效果。

令人意想不到的实用性元素不断渗入这一奢华住所。不常用到的狭窄的通道空间，戏谑性地用不锈钢名牌作标记，给人以生活空间的假象。

主卧的空间设计目的是根据生活和工作需要而设计。泥土色调和木质地板与新式家具浑然一体。放置在壁橱旁边的化妆台，有利于业主在自然光下进行阅读，或只是静坐观赏窗外风景。内置衣橱实现了室内空间的实际可用性。卧室具有典型的极简抽象艺术风格和实用性。漂亮整齐的白色线条有助于保持室内的整洁设计，而床下或内置衣橱都可以开发利用为存储空间。暖色调的墙壁色彩及木质地板都赋予了空间一种舒适感受，打造了有利于夜间睡眠的卧房空间环境。

Entering into the lift lobby, the first thing you can see is the stainless steel baffle beside which is a shoe cabinet with mirror. When you experience the warmth of wood flooring, and the laser cutting wood painting, you can feel not only a modern residence, but also an unique luxurious and comfortable homes. Restaurant reveals classical and modern mix styles. The ceiling is decorated with Tom Dixon designed lamps, light shines the white ecological wallpaper, just like the sun penetrating the jungle foliage, creating a effect of modern home in lush jungle.

Unexpected practicality elements continue to infiltrate into the luxurious residence. Not commonly used narrow channel space is branded with stainless steel brands, giving people a kind of illusion.

Space design purpose of the main bedroom is based on life and work needs. Earth tone and wooden floor and modern furniture are combined together. The dressing table beside the closet is good for the owners' read, or just sit to watch the scenery outside the window. Built-in wardrobe realizes the actual availability of interior space. Bedroom has a typical minimalist style and practicality. Pretty neat white lines help to keep the room clean. Space under the bed and inner closet can be utilized as storage space. Warm wall colors and wood floors give the space a kind of comfort, creating a bedroom environment that conducive to sleep.

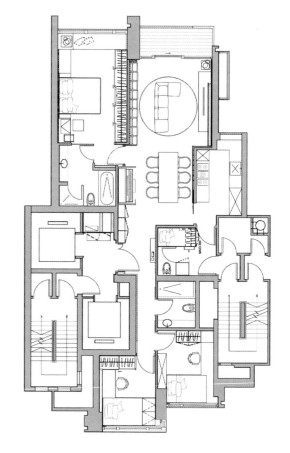

# Clovelly Court
## 嘉富丽苑

设 计 单 位：ANSON CHENG INTERIOR DESIGN LTD.
设 计 师：郑秋基
项目面积：204 ㎡
项目地点：香港
主要材料：雅士白大理石地台、胡桃木饰面、灰镜、黑油底磨砂玻璃

Design Company: ANSON CHENG INTERIOR DESIGN LTD.
Designer: Anson Cheng
Project Area: 204 m²
Project Location: Hong Kong
Major Materials: Aston white marble floor, Walnut veneer, Gray mirror, Black oil at the end of frosted glass

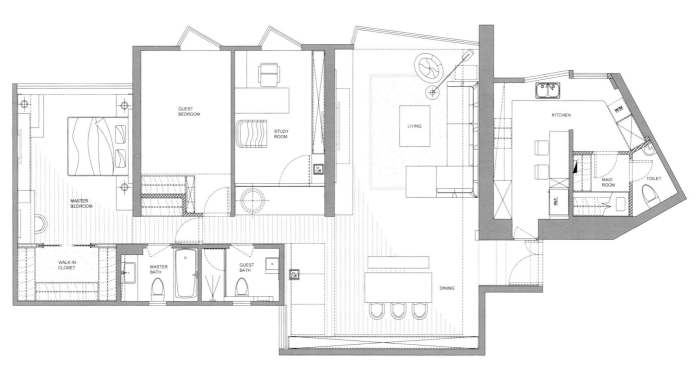

设计师利用简洁的几何线条、灯光和色调的配合，使简约的设计更显层次感和时尚感。客厅布置以简约为主，浅绿饰面板和天然黑石板组成的电视背景墙是焦点所在。特色地灯也点缀不少。室内有一面绿墙，恰与电视背景墙相互呼应。

餐厅设计同样使用了简洁的线条，餐厅中木餐桌和餐椅的设计，令空间充满大自然的感觉，设计师刻意用吊灯营造对比。餐桌后的一列黑色组柜与相邻的白色饰面墙构成鲜明对比，整体一气呵成。顶棚及墙身镜面的装饰令空间层次更显丰富。

Designer uses simple geometric lines, lights and colors' match, making the simple design has more layer and fashion sense. The living room is based on simple style. Light green veneer and natural black slate composes of TV background. The space has many featured lights. Indoor has a green wall, echoing with TV background.

Restaurant design also uses simple lines. Wooden dining table and chairs make the space full of natural feeling. Designer deliberately uses chandelier to create a contrast. A set of black cabinet and white decoration wall forms bright contrast. The overall design seems to be completed in one breath. Ceiling and wall mirror decoration makes room space more abundant.

# Laguna Verde Tower

海逸豪园

设计单位：深圳市林文学装饰工程有限公司
设 计 师：林文学
项目面积：105 ㎡
项目地点：香港
主要材料：实木复合地板、特色壁纸、黑镜、麻石砖

Design Company: (ShenZhen)Man Lam Interiors Design Co.,Ltd
Designer: Man Lam
Project Area: 105 m$^2$
Project Location: Hong Kong
Major Materials: Parquet, Special wallpaper, Black glass, Hemp brick

本案以简约手法来设计，室内以黑、白、灰为主色调，简单的衬托突显出每件家具、灯具等不同的艺术味道。由于基调偏冷，所以设计师用暖光增添了不少温暖的感觉，亦让整体节奏得以平衡。客、餐厅放了不少精致的摆设及画框，看起来很随意，但正是因为这些软装，给简洁的布置增添几分可爱。活动室用了特色报章壁纸，营造出另一番韵律，适合工作及阅读时的气氛。卧室以素色呈现眼前，与卫生间的粗犷感形成对比。在此，也想跟大家分享一下，很多业主喜欢在房子里堆满家具，而且在选材时习惯地选择一些多线条、多纹理的材料或家具，可能觉得色彩多变、线条繁复才有味道，才能活跃气氛。其实，这只是错觉，一个舒适的家就好像艺术馆，不需要有太多的堆放，只要放在里面的是珍品就好。空间应遵循"简约而不简单，简洁亦是美"的理念。

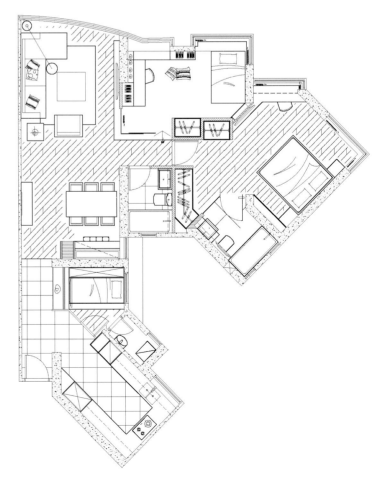

The case is designed by simple approach, interior space is mainly black, white and gray tones, the simple background highlights different artistic taste of every piece of furniture, lamps and so on. Since the tone is colder, designer adds a lot of warm feeling, and make the whole rhythm balanced. Living room, restaurant has many delicate furnishings and frame. They looks very casual, and they add a cute simple sense to the space. The activity room has many featured newspaper wallpaper, creating another rhythm. It is suitable for work and reading. Bedroom is decorated with plain color, forming a contrast with the toilet's rough sense. Here, I would like to share with you that many owners like to pile the house with furniture. They usually choose materials and furniture full of lines and textures, or they may identify colorful and complicated lines as good and lively. In fact, that is just an illusion, a comfortable home is like an art gallery. It is no need to have too much stack but one precious goods. The space should follow the concept "simple but not dull, simple is also beautiful".

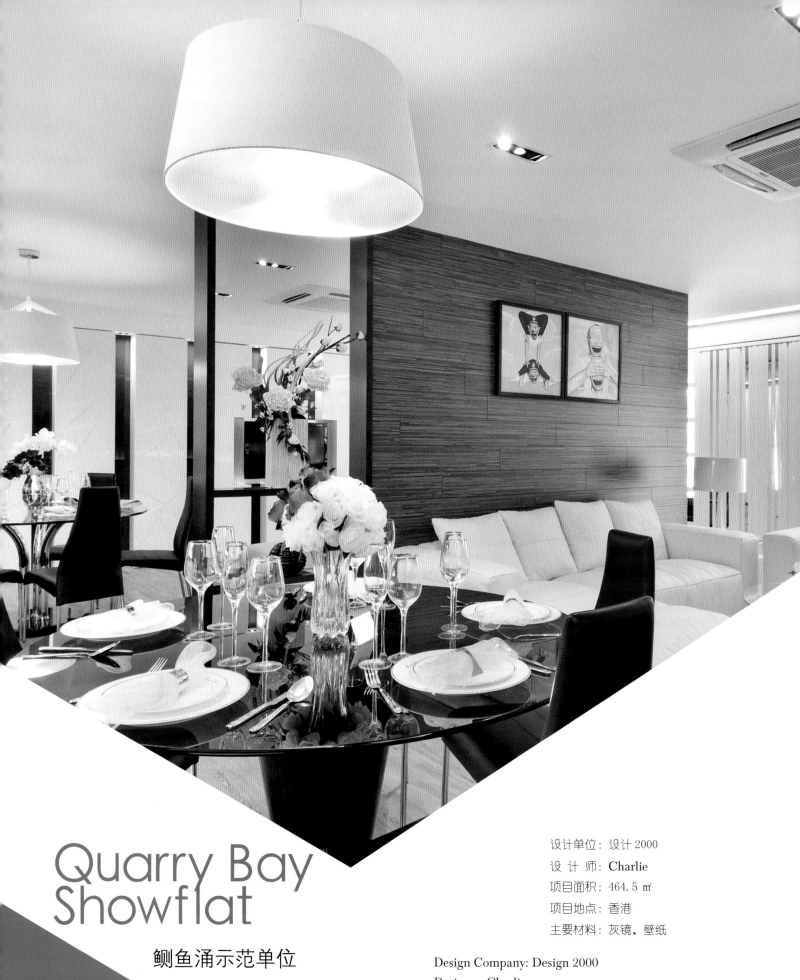

# Quarry Bay Showflat

鲗鱼涌示范单位

设计单位：设计 2000
设 计 师：Charlie
项目面积：464.5 ㎡
项目地点：香港
主要材料：灰镜、壁纸

Design Company: Design 2000
Designer: Charlie
Project Area: 464.5 m$^2$
Project Location: Hong Kong
Major Materials: Grey lens, Wallpaper

**主卧室及衣帽间**
主卧用带沉稳与亮泽的丝绒布艺作为装饰，令卧室气氛更显矜贵。设计师特意划出空间打造多功能卧房，为休憩区、衣帽间及卧室区作了流畅、实用的动线。黑玻璃的衣帽间在布置瑰丽的卧室中仿佛像一粒黑珠般闪亮。

**大地色系与黑白灰色系卧房**
米白色和深浅不同的啡色是打造自然风格卧房必备的元素，当中鲜明的木纹与深啡色壁纸为空间增添自然风情。把床铺设在升高地台中间位置，营造出随意的睡意气氛，一旁吊柜的门板焗灰黑色油漆，再镶上方格灰镜，使倒影成为特色装饰。

**特色公主房、少年房及儿童房**
七色梯级供上下床使用，床下设有衣橱，靠墙打造书桌，让面积有限的儿童房的功能齐全。粉红色搭配白色轻纱的床头布置，加上带利落花线的白色衣橱和床铺，展现出十分浪漫的公主房。黑白灰色系的卧房显然以男孩子为使用对象，椅子的线条也较为硬朗，吊灯选用了绅士帽造型。

Main bedroom and cloakroom

Main bedroom uses stable and bright velvet fabric as decoration, makes the bedroom's atmosphere more precious. Designer deliberately divides the space into multi-functional bedroom, adding resting area, cloakroom and bedroom smooth and practical generatrix. The cloakroom with black window is like a black pearl in the beautiful decorated bedroom.

Earth tone and black, white and gray tone bedroom

Beige and different brown color is the essential ingredient to create natural style bedroom, in which distinctive wood texture and dark coffee wallpaper adds natural style to the space. Laying the bed at the central of elevated platform, designer create a casual atmosphere. The side cabinet door uses baked gray and black paint, the drilled grid gray mirror makes the reflection becomes a featured decoration.

Unique angel room, youth room and kid room

Rainbow ladder is for getting in and out bed, under the bed there is a wardrobe. Beside the wall there is a desk, so that the limited area of the kid room has full function. Pink color and white veil bedhead arrangement and white wardrobe and bed show the romantic princess room. Black, white and gray bedroom apparently targets boys, chair's line is harder. Chandelier is gentleman hat shape.

# The Dynasty

御凯

设计单位：K's Creation Co.
设 计 师：陈伟建
项目面积：161.5 ㎡
项目地点：香港
主要材料：米白色大理石、
意大利仿木纹瓷砖、黑镜

Design Company: K's Creation Co.
Designer: Kenneth Chan
Project Area: 161.5 m$^2$
Project Location: Hong Kong
Major Materials: Beige marble, Italy imitation wood graintile, Black mirror

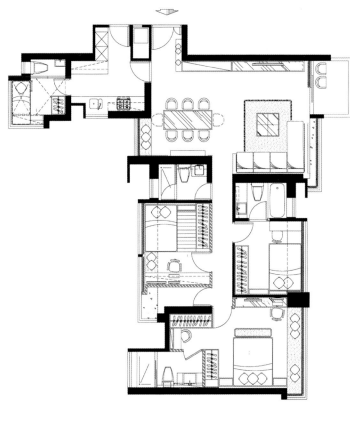

住宅的设计重点在于把握颜色的比重,设计师认为,深色调有点睛之效,但不可过度使用。因此他以大量浅色作为调和,减轻深色带来的压迫感。客厅电视背景墙铺设了不同厚度的米色大理石,形成良好的视觉效果。黑色电视柜如相框一般,强调了电视背景墙的美感。米黄色大沙发方便好客的业主接待朋友,L形的设计也方便业主一面躺卧、一面欣赏窗外的青马大桥美景。意大利米色仿木纹瓷砖中和了胡桃木纹沙发背景墙的硬朗感觉,为客厅注入了暖意。

设计师深谙特色墙设计之道,因而在空间中以不同的特色墙营造焦点。餐厅的特色墙与客厅电视背景墙一脉相承,黑色镜墙中巧妙地加入米白色长方形大理石,对比鲜明,能轻易将焦点引到浅色区域和大理石前面的独特造型雕塑上。镜墙前的黑色玻璃餐台和皮质餐椅也贯彻了这种对比,风格统一。

摒弃了简朴的卧室设计,设计师按不同家庭成员的品位,将几间卧室打造成各有特色的空间。主人房以祥和、沉稳的深啡色为主,床头背景墙用菱形扣布设计,提升了空间的格调。长条形黑镜装饰与床头灯槽均强调了修长的线条美,在视觉上拉长了房间的宽度。

男孩房的设计大胆、鲜明。设计师加设地台,把窗台变成床的一部分,大大地节省了空间。一片片木纹饰板悬于顶棚,不但可修饰横梁,还与纹理分明的深灰木纹壁纸和木地台相映成趣,令人耳目一新,展现出年轻人的活力。设计师充分考虑了点、线、面的空间运用,并通过特色墙营造出不同的感受,为业主带来有别于宫廷式设计的新尝试。

The design point lies in the portion of color. Designer thinks that the dark color has outstanding effect but can not be over used. So that he uses lots of light color as balance, reducing the pressure of dark color. Living room TV background has different level of beige marble, forming good visual effect. Black TV cabinet is like a frame, emphasizing the beauty of TV background. Friends may sit on the beige large sofa, the owner may sit on the L-shape design while enjoying the outdoor Qing Ma Bridge. Italian beige wood tile weakens the hard feel of hickory sofa background wall, adding warm sense to the living room.

Designer know well about the way of design, so he creates focus in the space on different characteristics. Restaurant's feature wall and living room's TV background wall have same origin. Black mirror wall has been added with white rectangle marble, the contrast is distinct. The focus has been led to the unique sculpture of in front of the light color area and the marble. The black glass stage and leather chair in front of the mirror wall carry out the contrast, forming similar style.

Designer abandons simple bedroom design, creating the room into different unique spaces according to the family members' tastes. The master room has peaceful and calm dark coffee, the

background wall has been covered by rhombus cloth, enhancing the space's style. Rectangle black mirror decoration and bed light groove emphasizes slim line beauty, lengthening the room's wide.

Boy room's design is bold and featured. Designer makes the floor stage, changing the windowsill to a part of the bed, largely saving the space. Pieces of wood panel is hanging on the ceiling, it is not only can decorate the crossbeam but also matches very well with the textured dark gray wallpaper and wood floor stage. People find it fresh and new, showing young people's vitality. Designer fully considers the space use of point, line and surface. He creates different feeling through characteristic wall, brings the owner a new try of lord design.

# The Astoria Tower
雅士花园

设计单位：深圳市林文学装饰工程有限公司
设 计 师：林文学
项目面积：225 ㎡
项目地点：香港旺角
主要材料：实木复合地板、特制壁纸、白色砖、火烧面文化石

Design Company: (ShenZhen)Man Lam Interiors Design Co.,Ltd.
Designer: Man Lam
Project Area: 225 m$^2$
Project Location: Mongkok, Hong Kong
Major Materials: Parquet, Special wallpaper, White brick, Flamed culture stone

用时尚与古典的完美结合来形容本案的设计是最适合的。入户的过道，大面积的镜子与亚光白墙互相呼应，加上木雕式的矮椅子及一幅带有乡村风俗的挂画，这种强烈对比的组合，在这套房子的每个角落，比比皆是，却又毫不突兀。树干的底座，玻璃台面的餐桌，配复古中式餐柜。单杆式悬浮楼梯，四周都是业主深爱的雅致摆设。布艺沙发当中，放着一张用原木掏空的茶几。到了二楼，更有铜钟、木雕、水墨画，被影楼式的灯具照射着，就连卫生间及户外部分，都同样精致。水滴式水龙头、梯子造型的毛巾架、20世纪70年代款式的电风扇，从每一件中都看得出设计师大胆的革新精神及深厚的搭配功力。

The captioned case is properly to be described as the combination of fashion and classics. The aisle, the mirror and the matt white walls are echoing with each other. There is a country style painting on the wall, and a wood carving chair in the room. The combination of those things forms a strong contrast in every corner of this house. The arrangement is suitable and not abrupt. The trunk base, glass table tops are matched vintage Chinese sideboard. Single rod suspended staircase is surrounded by the elegant furnishings beloved by the owner. Beside the fabric sofa there is a hollowed-out logs coffee table. Up to the second floor, there are bronze bell, wood carving, ink painting studio-style shining lamps; even toilet and outdoor part are equally fine. From drop taps, ladder shape towel rack or 70s style fans, one can see designer's bold innovation spirit and deep match skill.

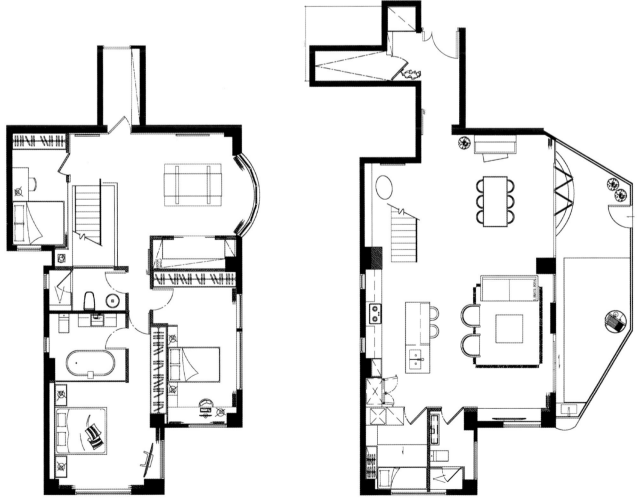

301

# Marina Cove

西贡匡湖居

设计单位：Millimeter Interior Design LTD.
设 计 师：廖子扬
项目面积：353 ㎡
项目地点：香港
主要材料：玻璃、大理石砖、水泥板、石纹砖

Design Company: Millimeter Interior Design LTD.
Designer: Michael Liu
Project Area: 353 m²
Project Location: Hong Kong
Major Materials: Glass, Marble tiles, Cement, Brick Shiwen

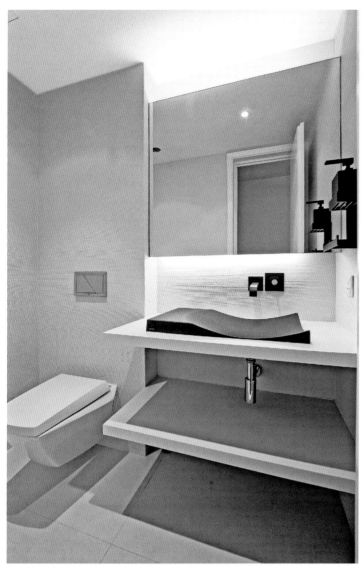

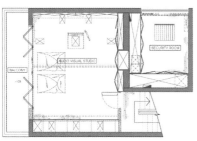

业主是第二次与设计师合作，故对其非常信任。业主的要求非常简单：希望住宅内设有车库、泳池及专业影音室。了解业主的期望后，设计师的设计方向是善用室外天然环境，可与室内相融合。利用住宅原有的建筑结构及特征，运用多样化的几何造型、比例及线条重新打造住宅空间。

设计方案制订后，设计师利用1∶1制作的3D效果图，向业主阐明设计造型、材料及采光等，令业主确切明白整体效果。业主没有提出任何改动意见，设计师于是用了7个月时间来改造。

此栋353 ㎡的独立屋共有6层，每层的设计各具特色。整栋屋子的设计不仅丰富，还依照了整体的设计概念，有一气呵成之感。低层划分为玄关位及车房，设计师特别设计2.3 m柚木大门用以平衡整栋房子的外形。地下是客厅，玻璃门打开后可把无边泳池及海景与室内融为一体。

一楼是厨房及餐厅，设计师采用落地窗，大大提升室内的视觉效果，原来较低及昏暗的空间顿时变得豁然开朗。

二楼是影音室，设计师采用层叠式设计，具有吸音及防止声音反弹的作用。再配以9.2环回立体声、3D投射器及150英寸透声投影幕，器材比一般影院更先进，专业程度无可比拟。

三楼是女儿房，房内有门通往浴室。浴室以白色大理石为主，散发出纯洁而高贵的气质。浴室与衣帽间互通并以清玻璃趟门作为间隔，方便使用者在同一空间梳洗整理。浴室的地台里铺装了S形热水铜喉，利用盛余的热能，使冬天在用热水时便可同时享有地暖作用，以达到善用能源及环保的效果。

四楼是卧室，房内以落地玻璃分隔浴室，使卧室在视觉上更为宽阔。浴室使用横条型炭灰色麻石衬托白色特色洁具，地台像三楼浴室一样铺装了S形热水铜喉，冬天时可享地暖效果。原来露台的位置特意改为放置按摩浴缸，置身其中可观赏180°海景。浴缸旁边装有电砂玻璃，需要隐蔽时可通电使清玻璃变为磨砂玻璃。设计师更将原有的顶棚打开，让日光充分进入屋内，晚上亦可观看夜空，令使用者置身于完全放松的空间。

设计师利用全高柜作为衣帽间及主卧的间隔。衣帽间的地台上特别设置一道小门通往监控室，监控室内设置了防盗系统、独立电话等，万一遇上盗窃，业主便可躲在里面等待救援。五楼的阁楼设置为书房，设计师刻意安排其远离二楼的影音室，令使用者可享受宁静的环境。

此住宅的设计时尚且散发着简约的美感，为业主创造了一个美观和实用集于一身的家。

This is the second collaboration between the owner and designer, so he trusts the designer very well. Owner's request is very simple: He hopes the house has a garage, a swimming pool and a professional audio-visual room. After understanding the owner's expectation, designer's design direction is using outdoor natural environment combining with indoor, using original residential structure and features and variety of geometric shapes, proportions and lines to re-create residential space.

After making sure the design, designer uses 1：1 3D drawing clearly states design modeling, materials, lighting and so on, so that the owner will fully understand the exact overall effect. Owner does not aim to make any changes, designer then uses seven months to transform it.

This 353 $m^2$ building is of six floors, each design is unique. The entire house is rich designed, forming an overall design concept and a sense of smooth. The lower floor is positioned as entrance place and garages. Designer specially designs 2.3 meters of teak door to balance the whole exterior of the house. The underground floor is positioned as living room, when opening the glass

door one can see the whole indoor infinity pool and sea views and interior space.

The first floor is kitchen and dining room. Designer uses french windows to greatly enhance the visual effect of interior. The original low and dim space suddenly become bright.

The second floor is audio-visual room, designer uses layer style design, which has an effect of acoustic absorption and reflection. The 9.2 surround sound, 3D projectors and 150 inches acoustically transparent projection screen are more advanced than other theater, the professional degree is unparalleled.

The third floor is the daughter room. The room has a door to the bathroom. Bathroom has white marble, exuding pure and noble qualities. Bathroom and cloakroom are exchanging with each other and the clear glass sliding door is a partition, therefore, user can groom in the same space. The bathroom floor has S-shaped copper water pipes. The surplus heat are used in winter to maintain warm water and floor heating to achieve efficient use of energy and environmental effects.

The fourth floor is the bedroom, the house uses floor glass to separate bathroom, the room seems wider in vision. Bathroom uses bar-type charcoal gray granite sets off white characteristic tools. Floor stage is installed S-shaped copper hot water jets likes the third floor bathroom. In winter user can enjoy warm floor effect. The original location of the terrace is changed into massage bathtub in which one can enjoy 180 degree sea view. Next to the bathtub is equipped with electric sand glass. The electricity can make the clear glass into abrazine glass. The designer open the original ceiling, so that the sunlight can be sufficiently into the house. At night, one can view the sky or being in a totally relaxed space.

Designer uses full-height cabinet as the interval of cloakroom and master bedroom. Cloakroom's floor stage has a small door leading to the control room where has security systems, independent telephone,etc., if the owner counters theft, he can hide inside and wait for rescue. Fifth floor's attic is a study, designer deliberately arranges it far from the second floor's audio-visual room, so that user can enjoy peaceful surroundings.

This house design is stylish and exudes a minimalist aesthetic, creating a beautiful and practical home for owner.

# Constellation Cove

涤涛山

设计单位：ANSON CHENG INTERIOR DESIGN LTD.
设 计 师：郑秋基
项目面积：353 ㎡
项目地点：香港
主要材料：黑色地毯、橡木地板、黑色木纹地柜、亚白色瓷砖

Design Company: ANSON CHENG INTERIOR DESIGN LTD.
Designer: Anson Cheng
Project Area: 353 m²
Project Location: Hong Kong
Major Materials: Black carpets, Oak floors, Black wood cabinets, Matte white tile

设计师的主要设计思路是在重塑各层空间、提升居室的格调之余，令空间更显宽敞。

本案是一个3层的独立屋，考虑到玄关面积过小，设计师拆掉了分隔玄关与休息间的墙身，令玄关的空间增大。客厅以几何线条与素净的白色调营造简约的感受，水泥电视背景墙的粗糙质感与空间的白色形成鲜明对比。设计师用了黑色线条勾勒出不同轮廓，丰富了整个空间层次。

厨房改为开放式，令空间更宽敞。餐厅与厨房皆以富有自然气息的材料点缀，前者用了木制的餐柜椅，后者则辅以石纹砖，延续贯穿室内的简约风格。

多功能室与餐厅以一玻璃门相隔，使空间层次更丰富。餐厅和多功能室与户外走廊相连，形成了开放的空间格局。位于上层的主卧较宽敞，于是将它分为三个区域，分别是睡眠区、工作区及衣帽间，房间也采用了简约的设计风格。

The main design idea of the designer is reshaping the space layers and space style, to make room more spacious.

The case is a three-floor individual house, in consideration of the small entrance area, the designer removes the partition wall between the entrance and the rest room, so that the entrance of space increases. Living room with white geometric lines and sober white tone create a simple feeling, cement TV background wall's rough texture and white space forms sharp contrast. Designer uses black lines to sketch out different outlines, enriching the space layers.

Open kitchen style makes the space wider. The restaurant and kitchen are full of natural materials. The former uses wooden sideboard, the latter is supported with stone texture brick, continuing the simple style throughout the interior space.

Functional rooms and restaurant are separated with a glass door, making the room's layer richer. Restaurant and functional room are connected with exterior corridor, forming an open space pattern. The main bedroom on the upper floor is wider, so it is divided into three regions, namely, sleeping area, work area and cloakroom, rooms design are simple style.

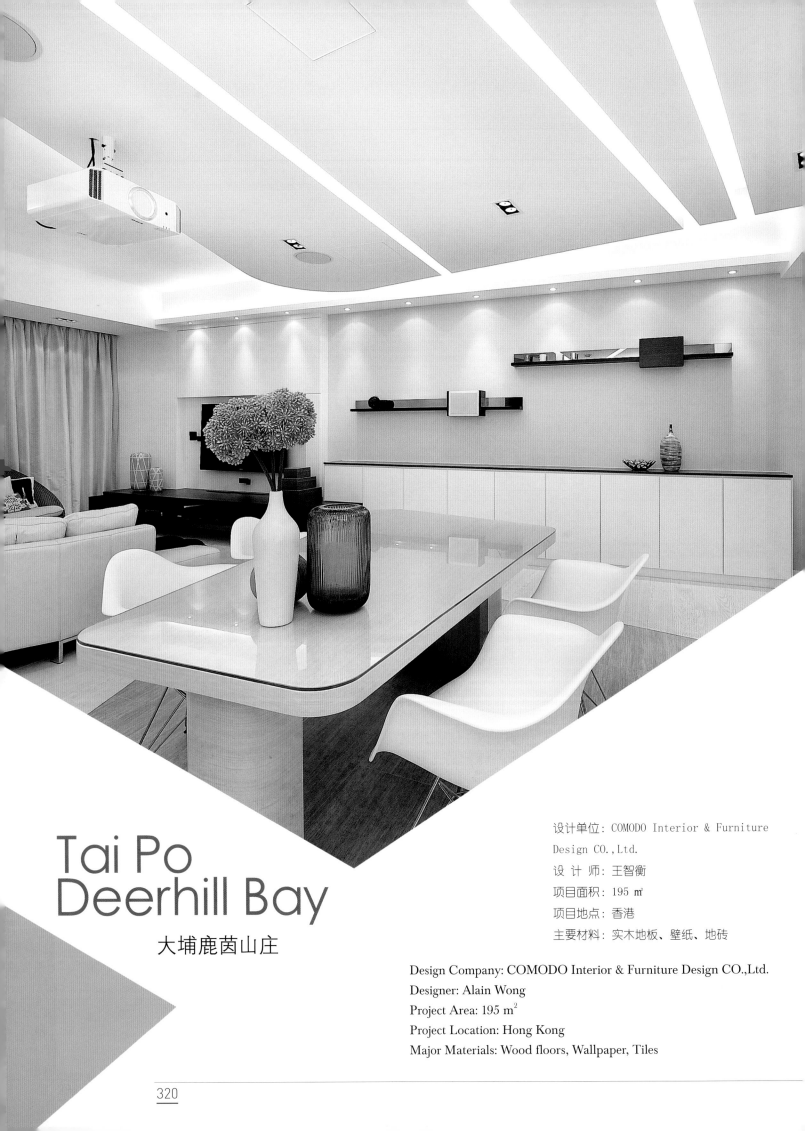

# Tai Po Deerhill Bay

大埔鹿茵山庄

设计单位：COMODO Interior & Furniture Design CO.,Ltd.
设 计 师：王智衡
项目面积：195 ㎡
项目地点：香港
主要材料：实木地板、壁纸、地砖

Design Company: COMODO Interior & Furniture Design CO.,Ltd.
Designer: Alain Wong
Project Area: 195 m$^2$
Project Location: Hong Kong
Major Materials: Wood floors, Wallpaper, Tiles

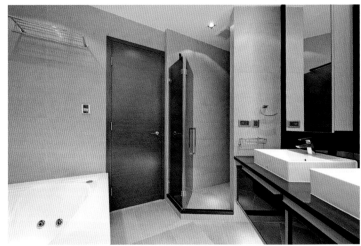

客厅、餐厅以浅色为基础，采用米白色实木地板，配以黑色及灰色的家具和材料，如深灰木饰柜面、淡灰色壁纸、灰色木皮特色墙身及浅灰色地台砖等。冷暖色交织将两厅表现得别具层次，而浅色色调加上触感丰富的材料，使得现代感和舒适感同时充斥着整个空间。

客、餐厅的亮点在于餐厅的不规则顶棚灯槽，由顶棚延伸至餐桌旁的灰色木皮特色墙，而地台铺砌的灰色地砖仿如顶棚灯槽的倒影，在丰富视觉效果之余，更突显了业主的个性及品位。而功能上，由于空间净高足够，顶棚稍微降低也不觉压迫，灯槽渗光有助于照亮空间，产生视觉开阔的效果，同时隐藏了冷气设备。

设计师亦考虑到空间的比例，客厅和餐厅应视为同一个空间来考虑。客厅用简单线条来营造层次感，电视柜及储物柜组合采用低矮的设计，在视觉上将空间拉长了，使整体的观感更轻盈连贯，亦将人们的目光集中在餐厅的顶棚灯槽上。黑白分明的开放式厨房充满现代感，具有光泽的灰玻璃墙身与白油玻璃互相辉映，线条简单的水晶吊灯看起来非常悦目。

设计师认为气氛会影响睡眠质量，而较深的暖色调有助睡眠，灯光的设计也不需要太亮。因此，将两房打通而成的主人房设计以较沉的色调为主，铺上深啡色实木地板、壁纸及木饰面，营造出温暖和温馨的气氛。睡床两侧是通往衣帽间及浴室的入口，而房间的另一端是工作的空间。浴室主要以黑、白、灰三色配搭触感丰富的地砖及木纹元素，营造出舒适但又富有现代感的格调。走廊地上装设了几个小地灯，光线由下而上地打在墙身上，令气氛更加柔和、舒适。

Living room and restaurant are based on light color, such as beige wood floor, black and gray furniture and materials, dark gray wood counter, light gray wallpaper, gray wood walls and light gray floor brick. Warm and cold colors make the two halls unique, light color and a touch of rich materials adds modern and comfort sense to the entire space.

Highlights of the living room and the restaurant is the irregular ceiling light groove, which extends from the ceiling to the gray wood wall. Gray floor tiles are like the reflection of light groove, apart from enriching visual effect, stressing the owner's personality and taste. As for the function, due to the space has enough height, the ceiling will not be oppressed if lower a little. Light groove penetration light helps to illuminate the space, forming open visual effects, while hiding the air-conditioning equipment.

Designer also thinks about the ratio of the space. Living and dining rooms should be considered as one space. Living room uses simple lines to create a sense of layers, TV cabinets and lockers use low design, and visually lengthen the space, so that the overall looking and feeling is lighter, people's attention will also be fixed on the restaurant's ceiling light groove. Black and white open kitchen is full of modern sense. Shiny gray glass walls and white oil glass are echoing with each other, simple lines crystal chandelier looks very beautiful.

Designer thinks the atmosphere will affect the quality of sleep, and deep warm color helps sleep, lighting design will not be too bright. Therefore, the master bedroom combined by two rooms is based on dark color. Dark coffee natural wood, wallpaper and wood finishes create a warm and welcoming atmosphere. Both sides of the bed are entrance to the cloakroom and bathroom, while the other end of the room is a work space. Bathroom is mainly in black, white, gray colors with brick and wood elements, creating a comfortable but very contemporary style. The corridor on the ground has been installed a few small lights, the light hits the wall from the bottom to the top, so the atmosphere is softer and more comfortable.

# Dynasty Heights I

帝景台（一）

设计单位：ANSON CHENG INTERIOR DESIGN LTD.
设 计 师：郑秋基
项目面积：246 ㎡
项目地点：香港
主要材料：人造绿地毯、木纹大理石、胡桃木地板、白色瓷砖

Design Company: ANSON CHENG INTERIOR DESIGN LTD.
Designer: Anson Cheng
Project Area: 246 m$^2$
Project Location: Hong Kong
Major Materials: Artificial green carpet, Wood marble, Walnut floors, White tile

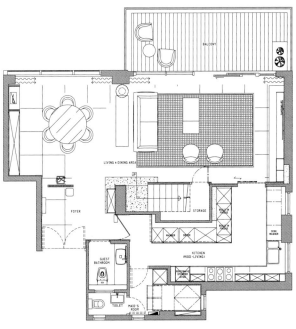

配上薄身设计的灰色皮灰沙发，质感柔和，另选配两张凳子，增加对比及趣味。

通往上层的楼梯地台以深色大理石为衬托，更显气派。楼梯空间采用黑色墙身并放上红色艺术品，营造出画廊的感觉。上层空间以房间为主，设计师考虑到业主一家人各自的颜色喜好，为各房间选配了不同的主色调，地面选用木地板以增加房间的温暖感。

The case is luxury style, different and carefully selected materials and reasonable layout create a luxurious and modern home environment. This is a duplex unit, the lower floor has restaurant and kitchen, the upper floor has main bedroom, children room and functional room. At the entrance, there are a dark coffee screen in the area. Interior space has ash wood marble. Designer uses different brown wood tones to create space level. The restaurant uses a simple line black glass round table and white chandeliers to add a luxurious atmosphere to the interior. Living room is covered with furry gray carpet. The gray leather sofa has soft texture. Two stools increase contrast and interest.

本案以奢华风格为主，透过精心挑选不同材料及合理的布局，去营造一种豪华而摩登的家居环境。这是一个复式单位，下层为餐厅及厨房，上层为主卧室、儿童房及多功能室。玄关处有一扇深啡色木条组成的屏风把区域分割。室内铺设了灰木纹大理石，也通过不同啡色调的木材来营造空间层次。餐厅选用了一张线条简约的黑玻璃圆桌，配上白色大吊灯，以增添室内的豪华气派。客厅铺上灰毛地毯，

The stairs leading to the upper floor is stylish dark marble. Stair space adopts black wall and red artwork, creating a gallery feeling. The upper space is mainly rooms, in consideration of the color preferences of the owner family, designer selects different color themes for each room. The ground is covered with wooden floor to increase warmth of the room.

## 图书在版编目（CIP）数据

名宅生活美学：时尚与舒适的空间对话 / 精品文化 编 . —武汉：华中科技大学出版社，2015.9
ISBN 978-7-5680-1026-9

Ⅰ.①名... Ⅱ.①精... Ⅲ.①住宅－室内装饰设计－图集 Ⅳ.① TU241-64

中国版本图书馆 CIP 数据核字 (2015) 第 157648 号

## 名宅生活美学 时尚与舒适的空间对话　　　　　　　精品文化 编

出版发行：华中科技大学出版社（中国·武汉）
地　　址：武汉市武昌珞喻路 1037 号（邮编：430074）
出 版 人：阮海洪

责任编辑：曾　晟　　　　　　　　　　　　　　　　　　　　　　　　责任监印：秦　英
责任校对：胡　雪　　　　　　　　　　　　　　　　　　　　　　　　美术编辑：张　艳

印　　刷：深圳当纳利印刷有限公司
开　　本：965 mm×1270 mm　1/16
印　　张：21
字　　数：302 千字
版　　次：2015 年 9 月第 1 版第 1 次印刷
定　　价：358 元 (USD 69.99)

投稿热线：(010)64155588-8000
本书若有印装质量问题，请向出版社营销中心调换
全国免费服务热线：400-6679-118 竭诚为您服务
版权所有　侵权必究